The Sustainable Development Goals in Higher Education

"Offering 'critical hope' for a more sustainable future, this clearly written book examines how higher education can engage in transforming itself and the world by considering the vision of social and environmental justice represented by the UN Sustainable Development Goals. Steele and Rickards challenge universities to become enablers of change by combining deep institutional commitments with bold, ethical cultures of innovation, and to integrate the SDGs into inclusive research and teaching. Especially impressive is the way in which the authors draw on social theory and constructive criticisms of higher education and development, to provide positive and actionable pathways forward. Reading the manuscript, I wanted to immediately share the insights of this book with my university leadership, my faculty colleagues and my students, especially as we try to navigate a post pandemic, antiracist, more equal and low carbon future for people and the planet."
—Diana Liverman, Regents Professor and Director of the *School of Geography, Development, and Environment, University of Arizona, USA*

"The world's universities are important for achieving the UN Sustainable Development Goals. What should universities be offering to this effort? Equally important, how should universities themselves change, in response to the SDGs? Wendy Steele and Lauren Rickards provide a unique, up-to-date guide to these issues, offering frameworks for thought and resources for action."
—Raewyn Connell, Emerita Professor, *Faculty of Arts and Social Sciences, University of Sydney*, and author of *The Good University*

"This book presents a compelling and refreshing take on the contemporary university and its transformative potential. Analysing the engagement between the Sustainable Development Goals and higher education, Steele and Rickards provide energy and hope for those of us exhausted by the continual restructuring of the neoliberal university."
—Lesley Head, Redmond Barry Distinguished Professor, *School of Geography, The University of Melbourne, Australia*

Wendy Steele • Lauren Rickards

The Sustainable Development Goals in Higher Education

A Transformative Agenda?

Wendy Steele
Centre for Urban Research
RMIT University
Melbourne, VIC, Australia

Lauren Rickards
Centre for Urban Research
RMIT University
Melbourne, VIC, Australia

ISBN 978-3-030-73574-6   ISBN 978-3-030-73575-3 (eBook)
https://doi.org/10.1007/978-3-030-73575-3

© The Editor(s) (if applicable) and The Author(s), under exclusive licence to Springer Nature Switzerland AG 2021
This work is subject to copyright. All rights are solely and exclusively licensed by the Publisher, whether the whole or part of the material is concerned, specifically the rights of translation, reprinting, reuse of illustrations, recitation, broadcasting, reproduction on microfilms or in any other physical way, and transmission or information storage and retrieval, electronic adaptation, computer software, or by similar or dissimilar methodology now known or hereafter developed.
The use of general descriptive names, registered names, trademarks, service marks, etc. in this publication does not imply, even in the absence of a specific statement, that such names are exempt from the relevant protective laws and regulations and therefore free for general use.
The publisher, the authors and the editors are safe to assume that the advice and information in this book are believed to be true and accurate at the date of publication. Neither the publisher nor the authors or the editors give a warranty, expressed or implied, with respect to the material contained herein or for any errors or omissions that may have been made. The publisher remains neutral with regard to jurisdictional claims in published maps and institutional affiliations.

This Palgrave Macmillan imprint is published by the registered company Springer Nature Switzerland AG.
The registered company address is: Gewerbestrasse 11, 6330 Cham, Switzerland

Preface

How Did We Get Here?

How did the world get to the point of entering the multi-catastrophic Anthropocene? How did universities get to the point of being places of intense paradox? And how did we, as two Australian university-based academics who research the impacts and antecedents of the climate emergency from a critical social science perspective, get to the point of writing this book on the Sustainable Development Goals (SDGs) and the role of higher education?

All of These Things Are Interrelated

As academics based in a public university system under enormous growth and development pressures with scaffolding flow-on effects for people, place and planet, we are highly alert to the fact that the crisis in higher education mirrors the broader climate of crisis and change. Despite the rhetoric around sustainability, impact and engagement, there is too little urgency or action within the mainstream Academy or in society on the big societal issues such as climate change, biodiversity loss and deep inequalities.

Decoupling the 'idea of the university' from the developmentalities that have led to current global conditions of unsustainability has never been more needed. And yet it has never felt so hard. The issue is that they are inextricably linked—joined at the hip—with shared histories, trajectories and legacies that must be first recognised and understood in order to be meaningfully addressed. This inconvenient, uncomfortable truth utterly disrupts the idea that universities are above, beyond or outside the worlds they critique and serve. It exposes the notion that universities can choose to engage, or not, as a myth—the question is how, not if, they affect the sustainable development challenge.

Sustainable Development for Who, How, When and to What Effect?

Now more than ever it is clear that we cannot neglect our global context and local connections. The SDG agenda is not perfect, but it holds radical potential relative to what many universities currently do and offers a platform and prospects for transformative action and change. Universities are at the heart of sustainability from the inside out. The SDGs are a shared witness statement to the state of the planet, one that requires humility and responsibility, but also brings with it hope that *a more sustainable future is still possible.*

Melbourne, VIC, Australia

Wendy Steele
Lauren Rickards

Acknowledgements

We would like to acknowledge that we live and work on the unceded lands of the *Wurrundjeri* people of the Eastern Kulin Nations and as uninvited guests we pay our sincere respects to Ancestors and Elders past, present and emerging.

This book began with a small project that received the RMIT University Enabling Capability Platform (ECP) 'Big Ideas' seed funding in 2018. We would like to acknowledge and thank Swee Mak and the ECP Directors for their support and engagement with the project. In particular, our thanks to Bruce Willis, Emma Shortis, Robbie Guevera, Simon Feeney, Renzo-Mori Junior, Jago Dodson, Prashanti Mayfield, Billie Giles-Corti, Ralph Horne, Joana Correria, Kim Gordon, Andrew Butt, Libby Porter and Olga Kokshagina for their constructive suggestions and encouragement. A number of key people have been supportive of the project and made time for meetings and discussion around the innovative role of the SDGs at RMIT University, including Calum Drummond, Mark McMillan, Belinda Tynan, Peter West, Katherine Johnson, Peter Kelly, Usha Iyer-Raniga, Clare Russell, Riaan Lourens, David Downes, Sherman Young, Linda Stevenson, Brian Coffee, Kylie Porter, Karyn Bosomworth, Melanie Davern, John Handmer and Darryn McEvoy. The project has been greatly enriched by the many contributions of the research associates and assistants: Oli Moraes, Mette Hotker, Anthony Richardson, Jaydon Holmes, Sarah Robertson and Tom Overton-Skinner.

Externally we would like to thank Scott Rawlings (Office of the Commissioner for Environmental Sustainability), John Thwaites (Monash University), Councillor Cathy Oke (City of Melbourne), and Lars Coenen and Michele Acuto (University of Melbourne) for their early engagement and encouragement. Our thanks also to the University of Johannesburg, Association of Commonwealth Universities and British Council for their support and promotion of the SDGs and higher education theme, in particular Consola Evans, Alex Wright and Georgina Nicolini. At Palgrave Macmillan we sincerely thank Rachael Ballard and Joanna O'Neill for their support and encouragement.

Finally, we would both like to thank our families who understand all too well the pressures of juggling multiple roles during a pandemic lockdown. Wendy would like to thank Russell and Jesse for sharing the journey with such love, as well as Libby, Crystal, Karyn, Brian and Kristen for laughter and light in these uncertain times. Lauren would like to thank Gemma and Selby for their immense patience and inspiration, as well as Zoe, Leanne, Kim, Karyn, Susie and Andrew for their fierce support and encouragement.

Contents

1 A Transformative Agenda 1

2 Sustainable Development in the Anthropocene 35

3 The Role of the University in Society 67

4 Ethical Innovation 107

5 Re-thinking Research Engagement 145

6 Learning and Teaching Matters 169

7 What Does Success Look Like? 205

8 Sustainable Futures 247

Bibliography 269

Index 285

About the Authors

Wendy Steele is Associate Professor in Sustainability and Urban Planning in the Centre for Urban Research (CUR) at RMIT University, Melbourne, Australia. Her research focuses on cities in climate change and human-nature relationships in the urban age, as a key dimension of scholarship and education in, for, and about environmentally sustainable development. She is the author of six books on the role and impact of quiet activism, critical urban governance, climate justice and sustainability as a transformative agenda.

Lauren Rickards is Professor in the Centre for Urban Research (CUR) and Director of Urban Futures Enabling Capability Platform at RMIT University, Melbourne, Australia. Based in human geography, her research focuses on social responses to the climate change crisis, including the role of climate change in sustainable development, and implications for the research and education sectors.

xi

List of Figures

Fig. 1.1	Universities and the SDGs: a transformative agenda. (Source: Authors)	6
Fig. 1.2	Four possible scenarios for university engagement with the SDGs. (Source: Authors)	19
Fig. 2.1	The four worldviews of Cultural Theory. (Adapted from the work of Mary Douglas and colleagues. Figure from https://www.dustinstoltz.com/blog/2014/06/04/diagram-of-theory-douglas-and-wildavskys-gridgroup-typology-of-worldviews)	50
Fig. 3.1	Embedded and engaged—expectations of twenty-first-century universities (Source: Authors)	71
Fig. 3.2	The civic university. (From Goddard 2018, p. 263)	92
Fig. 4.1	Leverage points for systems change. (From Fischer, J., Riechers, M. (2019) A leverage points perspective on sustainability. *People and Nature* 1, p. 115–120)	134
Fig. 6.1	The sustainable development triangle	173
Fig. 6.2	Sustainable Development Goals 2030 (SDGs) wedding cake. (Source: Rockström and Sukhdev, Stockholm Resilience Centre, 2016)	174
Fig. 6.3	Three critical frames for SDGs learning and teaching. (Source: Authors)	176
Fig. 7.1	Universities and the SDGs—past, present and future	207
Fig. 7.2	Four possible scenarios for university engagement with the SDGs, overlaid with the four worldviews of Cultural Theory	210

Fig. 7.3 Existing and emerging roles for libraries and the SDGs. (Adapted from IFLA 2020) 238

Fig. 8.1 The Three Horizons framework. (Adaptation by the social enterprise, The H3 Uni https://www.h3uni.org/practices/foresight-three-horizons/) 259

Fig. 8.2 Re-imagining the future of the university—six priorities. (Adapted from Bina and Pereira 2020, p. 22) 261

List of Tables

Table 2.1	The 17 SDGs and the 8 MDGs	37
Table 4.1	The RADAR framework of ethical innovation	124
Table 6.1	Ethical pedagogical innovation	181
Table 6.2	SDG Skills and Competences	189
Table 7.1	Embedding the SDGs in higher education through an indicator framework—example of AQUA and ACPUA's governance and strategy framework	219
Table 7.2	Generating impact in higher education	229

1

A Transformative Agenda

Bearing Witness

> We have overrun the world …
> The real threat is not to the survival of the planet, but to the survival of humanity.[1]

These are the raw statements by ninety-four-year-old British filmmaker and historian David Attenborough in *A Life on our Planet*. For much of the film he stares directly into the camera as he describes the world's devastating biodiversity loss at the hands of humanity, the furious pace of human progress, unconstrained consumption of finite natural resources and the cumulative impacts of the climate emergency. This is his witness statement: 'a stark warning of how—as a society—we have squandered this gift'.[2] He remains hopeful however that a different, more sustainable future is possible.

That other sustainable worlds are still possible is similarly the central message in the decolonial manifestos of *Buen vivir* (South America), *Ubuntu* (South Africa) and *Swaraj* (India). These visions of social and ecological commons focus on futures that are community-centric, ecologically balanced and culturally sensitive. 'It's a vision and a platform for thinking and practising alternative futures focused on lived practice, that is aware of—and connected to—global movements of local solidarities that promote collaborative consumption and economies of sharing and care'.[3] The aim is to fundamentally *repoliticise* sustainability and its links to development trajectories. As the *Uluru Statement of the Heart* in Australia eloquently states:

> sovereignty is a spiritual notion: the ancestral tie between the land, or 'mother nature', and the Aboriginal and Torres Strait Islander peoples who were born therefrom, remain attached thereto, and must one day return thither to be united with our ancestors. This link is the basis of the ownership of the soil, or better, of sovereignty. It has never been ceded or extinguished and co-exists with the sovereignty of the Crown. How could it be otherwise? That peoples possessed a land for sixty millennia and this sacred link disappears from world history in merely the last two hundred years?[4]

At the heart of the idea of sustainable development are the prospects for future *sustainability* historically linked to the trajectory and legacy of modern capitalist *development*. In this sense sustainability and development sit 'against' each other. As Laura Kipnis describes, 'to be against' has multiple meanings[5]: it can be to stand opposed, but also to lean together or towards, foster and bolster.[6] It is within this relational context that we explore the role and contribution of the *Sustainable Development Goals* (SDGs) as a transformative framework within the context of higher education.

The key premise of this book is that new progressive directions and possibilities for deeply engaging with the SDGs are opening up for universities—and yet remain under threat. As a United Nations-led and goal-driven initiative, the SDG agenda is not without risks and, like universities, is rightly subjected to criticism about the inadequacy of 'the master's tools for dismantling the master's house'.[7] However as civil rights and feminist activist Audre Lorde goes on to say, 'in our world, divide and conquer must become define and empower'.

Our approach to the book is not to polarise the SDGs as 'sinner or saint', but instead to critically position them as an imperfect but crucial and collective witness statement to the unsustainability of our age. By focusing on the critical role of education *about, for* and *through* the SDGs, we seek to advance constructive engagement with higher education that is both progressive and meaningful.[8] We are all responsible for bearing witness to the ecocide and genocide being driven by unsustainable modern development (including in higher education) with its aggressive economic growth and an ongoing colonial legacy. In higher education and elsewhere, a transformative agenda is needed that addresses this unsustainability in ways that are genuine and regenerative. The SDGs offer a starting point for such work, if we shift the emphasis from 'cockpitism' to critical co-production in place and practice.[9]

To 'bear witness' is not a passive position, but instead offers a powerful way of working through difficulties or trauma by being both present and committed to critical, regenerative action. This involves the humility and empathy that 'moves individuals from the personal act of 'seeing' to the adoption of a public stance by which they become part of a collective, working through trauma together'.[10] Bearing witness means recognising collective responsibility for unsustainable development trajectories and impacts ('developmentalities') and using it to help move towards recovery— rather than just turning away from a painful past, or even towards a disconnected utopian future. This is 'not merely to narrate, but to commit oneself and the narrative to others: to take responsibility for history or for the truth of the occurrence'.[11] This is the starting point for collective action and healing.

In the current climate of environmental change and societal crisis, higher education needs to both engender and embrace this responsibility, humility and regenerative praxis. As the SDG agenda makes clear, universities are a key tool for implementing the SDGs. They are also far more than this. They are the products and perpetrators of the same growth developmentalities that continue to generate the Anthropocene, as well as expressions of the same progressive ambitions and traditions that animate the SDG agenda. These resonances between universities and the SDGs mean that higher education—complete with its ambiguities, tensions and potential—is ideally placed to proactively engage.

Universities have a unique capability to find, explore and translate progressive ideas; to seek and adapt new critical lenses; and develop creative ways of unsettling the world—including disrupting or re-formulating areas where ideas and action around the SDGs have and will become stuck. At least that is the theory. In practice, universities' capabilities are often severely constrained by the very sorts of issues that the SDGs draw attention to—issues such as inequalities, a lack of decent work, poor governance and vulnerability to disruptions. Combined with their far-reaching impact on the planet, this means that universities need the SDG agenda as much as the SDG agenda needs universities.

Although most discussion about the two is framed as 'how can universities contribute to the SDGs?', the contribution is two way. The idea that universities' role is to help others' address the SDGs reflects a deeply unhelpful presumption that universities are separate to the world the SDGs are addressing. From such a presumption flows the self-serving misconceptions that universities are mere observers of, not drivers of, the unsustainable condition of the world, and that they are free to choose *whether* or not they contribute to the SDG agenda rather than address how they already are (for better or worse) affecting the agenda and its prospects. The SDG-university relationship is one of co-production and the question is what role universities play. When rooted in honesty and humility, this role can involve forging 'new concepts and new productive ethical relations'.[12] It also needs to be about what Rosi Braidotti describes as:

> coming to terms with the unprecedented changes and transformations as the basic unit of reference of what counts to be human.[13]

Critical engagement with the SDGs in the sense we envision involves facing—not running from or brushing over—flaws in the SDG agenda and recognising that these flaws and their roots are shared by universities. It involves understanding the reciprocal role of universities within the contemporary sustainable development challenges presented by the Anthropocene, and heeding the SDG agenda's call to face unsustainability; boost resilience, adaptation and experimentation; and invest in maintenance, repair and regeneration. Such critical engagement helps address the inevitable question of 'what should we do?'. To this end, our key arguments that drive the book can be summarised as two-fold.

- *First*, that as an integrative, transformational agenda, the SDGs demand approaches that work across boundaries, and that connect efforts across different issues to identify synergies and tensions. For this reason, the SDG agenda is not just one among many topics or areas of work within a university, it is a framework and context that demands a new way of working in all aspects of universities. When it comes to the SDGs, universities are not just enablers of change but also *targets* of change.
- *Second*, that individually and collectively we are already engaging with the SDGs by virtue of being part of the world it represents. For all of us—including universities—the question is *not if, but how and in what ways* do we want to engage with the challenges and opportunities of sustainability in a climate of change? While maladaptive business as usual is possible, so too is a transformative approach involving deep institutional commitment and a bold, innovation culture as the pathways to sustainability-led change.

The SDGs as a transformative agenda can serve to bring universities 'back to e/Earth' by underlining that all of us and all institutions need to comprehensively change in order to get society onto a more sustainable and just pathway. This is about more than getting universities to more actively help others. It is about improving the consistency between what universities say and what they do, and closing the enormous gap between occasional references to 'sustainable development' in strategic plans or curricula, and the actual impact universities are currently having in the world.

Within both the University and society more broadly, the SDGs demand approaches that work better to scale up, out and deep[14] the local and international efforts that are needed to sustain *all* types of life on an increasingly warming and unequal planet. This includes working at the nexus of issues such as water, food, carbon, climate and health as a crosscutting interdisciplinary and multi-stakeholder agenda that links academia with the rest of the society. It involves not only new content and projects, but new structures, processes, cultural norms and ethos that enable universities to critically evaluate their role in (un)sustainable development and address their own ambiguities and paradoxes (see Fig. 1.1 below).

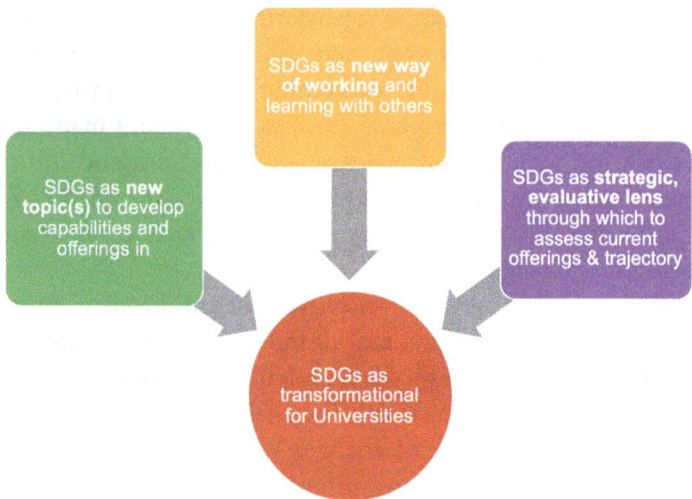

Fig. 1.1 Universities and the SDGs: a transformative agenda. (Source: Authors)

To be a transformative agenda, the SDGs must become embedded in everything universities do, including leadership, strategies, research, learning and teaching, partnerships, operations, advocacy and activism. The SDGs are not just one among many topic areas within a university, they are a strategic focus and context that demand a new way of working and offer political opportunities for addressing deep structural inequities. As Isaac Kamola highlights within the South African anti-apartheid context, while:

> universities imagine themselves as "global", settler colonialism and racial apartheid—and acts of resistance to them—continue to shape higher education. Efforts to engage this historical legacy can serve as a point of inspiration for those critical of the current state of higher education around the world … activists—both past and present—know that universities contain vast political possibilities and that part of reclaiming these possibilities requires demanding that the university be otherwise.[15]

In the following sections of this introductory chapter, we emphasise that transformative change is a reciprocal agenda that addresses both the monsters 'out there' as well as 'in here'. We outline the paradoxical role of both the SDGs and universities before turning to articulate our critical, social science approach inspired by feminist scholarship. To this end our focus is on the need for regenerative responses, ones that aim not only for the neutralisation of negatives but for the cultivation of new, positive possibilities. This is what we believe—in their best light—both SDGs and the universities offer as a transformative agenda. The SDGs prompt us to ask: what do we want to grow within universities, and what do we want to weed out in order to translate the agenda into a regenerative tool on the ground?

Facing Monsters

Transformative change is a *reciprocal* agenda which requires critically reflexive action and change both 'in here' (i.e. within the academy, Universities, higher education) and 'out there' where universities are entwined with and part of society more broadly. A global pandemic such as COVID-19, the more localised disasters of bushfires, droughts or floods, or global corporate and bureaucratic systems for example, can take on monstrous lives of their own, full of unimaginable horror.

The monsters we fear say a lot about ourselves and our society, our fetishes and our anxieties.[16] The monster metaphor has been used to describe multinational corporations and more broadly the growth of capitalism and economic ideologies which underpins them, from the fearsome Scandinavian sea monster 'The Kraken', to the blood-sucking Vampire, to Frankenstein and the Zombie walking, the living un-dead.[17] The 'Corporate Frankenstein Monster' is a descriptor of 'plundering, pillaging, and polluting the planet for profit'.[18] As anthropologist Hariz Halilovich observes from his research into forced displacement and diaspora in post-war Bosnia, what is really frightening is that the monstrosity apparent in many human activities is real.[19]

Some critical thinkers reject the SDGs as not radical enough, as yet another example of 'the masters' tools' that have generated our contemporary crises. The goals are read as just another 'developmentality'—or monster—in our midst: top-down, hierarchical, imperial by design and nature, driven by instrumental goals and indicators that are neglectful of people and place and in particular Indigenous cultures and localised places. The SDGs, it has been argued, threaten to further legitimise or reinforce the systems of injustice and lack of sustainability that define the neoliberalised development *status quo*.

Political ecologist Maria Kaika, for example, argues that despite the rhetoric of a 'paradigm shift' for pursuing the SDGs, the emphasis remains dependent on 'old methodological tools (e.g. indicators), techno-managerial solutions (e.g. smart cities), and institutional frameworks of an ecological modernization paradigm that did not work'.[20] She calls instead for agendas, frameworks and practices that serve to disrupt path dependency in order to establish alternative methods for achieving social equity and environmental sustainability that sit outside the current *status quo*. In particular she is concerned that the SDGs' emphasis focuses on 'what' needs to change, rather than 'how' this change can be achieved through different practices.

Another serious and legitimate critique of the SDG agenda is its lack of explicit recognition or engagement with Indigenous rights and sovereignty, especially given the agenda's stated commitment to ensuring that 'no one is left behind'. This omission is further highlighted by critics who argue that the application of the SDGs in universities: (1) serves to further an econo-centric approach to ecological and sustainability education that risks ameliorating other ways of knowing and learning, such as Indigenous ontologies and nature-based pedagogies[21]; and (2) that the focus on the SDGs in pedagogy and research can further entrench the neoliberalisation of the University and the ways in which sustainability pedagogy and research develop in higher education to 2030 and beyond.[22] There is a risk that capacities for systemic transformation are muted through homogenous development discourses that do not reflect local contexts, imposing knowledge from elsewhere in ways that erase local ways of knowing and doing.[23] This is 'the monster that constantly reshapes itself to haunt the culture that is using it—not just the culture that created it'.[24]

These warnings about and weaknesses in the SDG agenda need to be taken seriously and used as a constant reminder not to think of the agenda as some kind of magic formula. Some aspects of the SDG agenda are far too accepting of the existing context it has emerged out of. The whole agenda needs to be handled in a way that is fully cognisant of the risk that unthinkingly applying it may reproduce, not dismantle, structural problems and injustices. Empirical research has already documented, for instance, the ways in which the SDG agenda is being co-opted in some situations to reinforce not disassemble extractivist fossil fuel logics. How the SDG agenda plays out in practice is far from guaranteed.[25]

But these risks and monstrous aspects of the SDG agenda are exactly why academic engagement is needed. Furthermore, such engagement is needed because the academic context is characterised by the same challenges. Neither the SDG agenda, local initiatives in its name, nor universities are context- or problem-free. In our opinion, the resultant challenge is not to wait for a future perfect agenda, free of the taint of the current world and enacted without tension in diverse contexts, but to get started, knowing that scrutiny and difficult intellectual and political work are needed along the way. We say this as academics in Australia, where it has long been clear that we cannot wait for perfect plans from our political leaders, and instead need to be clever in subversively utilising what is available.

Critical, serious and mischievous engagement of the sort that academics are especially well positioned to foster is needed to help drive the SDG agenda *while* improving it and keeping it on track. As enablers *and* targets of change, universities are vital to the overall success of the SDG agenda as a 'living' transformative agenda and proliferating collection of positive initiatives. Academics and academic institutions can be powerful change agents on all levels of the agenda. They have the capacity to draw on in-depth analysis to highlight lessons from the past, interrogate the present, discern genuine opportunities and identify how—despite the risks—the SDG agenda could be truly transformative moving into the future.

The seriousness of the global challenges covered by the SDGs makes it imperative that higher education does not turn away from the SDG agenda. Rather, there is a need to help shape what the agenda means in practice and make it the transformative catalyst it needs to be. The 'regions of human practice with old or established boundaries are being challenged by new ensembles and configurations … and can reveal the origin, identity, purposes and powers of the monster, and in doing so, ourselves'.[26]

For us personally, the SDG agenda reminds us of the dangers not just of co-optation but of cynicism and perfectionism. While critique of ill-considered change agendas is essential, the monster we are most afraid of in the context of the horrors of the Anthropocene is the one that traps us in its web of criticisms, caveats and academic posturing. We need to act, and the SDG agenda helps us do so. That alone is reason to engage with it.

We appreciate the tensions and ironies in taking this stance. But irony is itself a tool for dealing with the challenges of the Anthropocene;[27] not in the sense of a postmodern 'dispositional irony' that 'freezes irony into an aesthetic pose'[28] and breeds cynicism, but in the sense of irony as 'among our best methods for immediately and unconsciously adjusting to complex circumstances' and coping with the disparities and 'inchoateness of the human condition'.[29] This is about an ironic relation to the world, one that appreciates that the world's inherent relationality means it always exceeds our understanding but also means we cannot help but act, even if (or perhaps especially if) we do nothing. Bronislaw Szerszynski argues that an 'ironic world relation' offers a way to both recognise 'failure and error' and push us 'to act, with due care, in the very face of that recognition'.[30]

In this way, embracing irony and imperfection helps us address 'the ecological paradox' of informed inaction[31] and the 'politics of actually existing unsustainability'[32] that characterises the role of universities in current (un)sustainable development. As we outline in this book, it calls on us to consider more deeply the implications of the SDG agenda for the university sector, and the implications of not waking up.

Who's Afraid of the SDGs?

The critiques being raised of the SDGs are vitally important to attend to as both the means and ends of our current planetary-scale crisis are deeply and inextricably linked to the prospects and possibilities for transformative change. These criticisms are also reflected in the critiques of the modern university: from its colonial origins through to neoliberal reform and many universities' prioritisation of profit over public service, financial return over investment, and performance indicators and net promoter scores over real 'impact'.

Just like the SDGs, 'the university' is characterised by complexity, tensions and inconsistencies that can serve to inspire or inhibit, impoverish or empower, hurt or heal. In particular, the university is a place of paradox that holds both conservative and transformational tendencies. There are at least three common manifestations of this paradox we would like to draw attention to.

1. *Tradition and radical change*—As institutions, universities and associated groups such as academic disciplines can be deeply resistant to change, which is one reason they have been both relatively immune to disruption over the centuries and repeatedly targeted for 'makeovers' by private sector interests. At the same time, the Academy and higher education is founded on a commitment to intellectual freedom and critique, a generator of novelty and innovation, and an enabler (if not always site) of profound social change.
2. *Wealth and financial precarity*—Universities have the ability to generate and concentrate both great wealth and great financial precarity. As COVID-19 and the related economic crisis have exposed starkly, some institutions, disciplinary areas and staff are disproportionately wealthy and secure, while financial and career precarity have become ever more thoroughly entrenched for others (notably casualised staff, many students, and universities outside the global elite).[33]
3. *Inclusion and exclusion*—As institutions committed to the value of ideas and knowledge as a common good, universities espouse and facilitate democracy and openness. Their relative independence means many universities can actively embed inclusive and democratic practices and try to promote and enable them in wider society, including by providing citizens with important insights and information. Yet universities also have the capacity to exclude, exploit and entrench concentrations of power and privilege. Whether manifest in who has access or whose voices are prioritised in curricula, partnerships and university decision-making, universities can be welcoming and open-minded, or hostile and oligarchic.

To address these contradictions and tensions, we must do more than just critique the SDG agenda as the new monster in our midst. Critique allows us to 'unveil, uncover and critically re-examine the convincing logics and operations' of truth claims. While useful in finding fault—and certainly a technique we use in this book—critique retains 'a certain external knowingness, a certain ability to look in from the outside and unravel and examine and expose that which had seemingly lay hidden'. It is thus insufficient in helping us address the world of global sustainable development and universities that we are part of, especially given that the current unsustainable state of the world points in myriad ways to the collapse of the dichotomies of 'inside' and 'outside'.[34] Critique is also liable to paranoia of the sort that sees only negatives.[35] To negotiate these challenges, we need not only irony but what Irit Rogoff calls 'criticality': 'a double occupation in which we are both fully armed with the knowledges of critique, able to analyse and unveil, while at the same time sharing and living out the very conditions which we are able to see through'.[36]

Both the SDGs and universities are complex, diverse assemblages of people, practices, materials, spaces, conversations, initiatives and ideas that have long historical roots and are continually shifting and remade every day. Their outcomes are necessarily experimental and intersecting. As William Mosely notes, the SDG agenda is one part of 'myriad [...] development experiments (or natural experiments) to try to improve the human experience' underway in the world. Universities have long been central to these experiments and remain so in the era of the SDGs, regardless of whether they acknowledge it.[37]

In the play *Who's Afraid of Virginia Wolf?* Edward Albee implicitly examines the relationship between universities and society. Ablee paints a portrait of a ruined Western civilisation balanced between history and science on the one hand, and the brutal relationship of university professor George and his partner Martha on the other. Set at an after party of university colleagues that descends into a 'boozy marital slug-fest', the play presents George and Martha tearing each other apart with word games that continually confound the difference between truth and illusion. George 'vacillates between detachment and involvement' in the nastiness he helps precipitate, including adopting the classic academic stance of detachment—that of a commentator on the chaos unfolding around and through him.[38] Written in the early 1960s when the US was emerging from the 'narcoleptic Eisenhower years when a fragile cold war peace that depended on the balance of terror',[39] the play presents the dysfunctional politics and monstrosity of middle-class American marriage, values and universities as a devastating microcosm of, and parable about, the dangers of self-delusion/destruction amidst the violence, complacency and excess of Western modernity.

Universities remain microcosms of wider society and its monstrous politics. Similarly, academics frequently 'vacillate between detachment and involvement' in how they attempt to relate to this broader context as well as their own more local ones. Thanks in part to the culture of heightened competition that now pervades universities, many academics ignore much of the world but invest large amounts of emotional and physical energy into brutal scholarly encounters in the Academy, striving to distinguish themselves by contesting others. This points us to a further danger of critique: that criticisms are driven by a habitual contrariness and desire for point scoring rather than a deep conviction that critique is actually productive in a given situation.

Critique clearly can be productive in terms of the SDGs, but it needs to stem from a commitment to engaging not merely with arguments but with consequences and outcomes. Geographer Diana Liverman, for instance, calls on geographers to engage more deeply and systematically with the SDGs in creative and constructive ways. Highly alert to the paradoxes and perversities of the SDGs, she calls out the paradoxes and perversities of academics refusing to engage with such a global agenda, particularly given the privileged capacity many of us have to 'work within the system'. She asks:

> Can we constructively engage with the post-2015 development agenda and the SDGs in ways that are progressive and meaningful? And what does constructive engagement imply for our everyday scholarship, service, and outreach?

Taking up Liverman's provocation, Farhana Sultana concurs:

If we want emancipatory politics and transformations in development, we need to challenge and improve what is done in the name of SDGs, keeping central the issues of social justice and ethical engagement. ... We need to reassess what it means for us to be 'engaged' scholars, and what kind of impact we hope for (whether achievable or not). [...] We need to engage critically and constructively, however we can. Too much is at stake to not do so. If the SDGs are truly to be useful and have transformative potential, then we must be part of that conversation too, and develop new tools to dismantle the master's house.[40]

When we use the term 'universities' or 'the university' in this book we do so merely as a shorthand for what we know is a highly heterogeneous and dynamic institution and sector. Indeed, it is the existence of diversity and change within the sector that fuels our argument that today's 'typical' university could be otherwise. We also use the term SDGs knowing what a messy array of ideas and voices they contain, and what varied interpretations and implementation efforts they are stimulating. Again, it is the internal heterogeneity and capacity for manoeuvre and co-production that we find one of the most interesting and motivating things about them.

It is because the SDG agenda and universities are not fixed or given that we believe the SDG-university relation should not be superficially decreed, nor rejected out of hand. Even within the constraints of heavily neoliberalised universities, there are innumerable opportunities for university staff and students to work in creative and critical ways with the SDGs. Even just beginning with a few of the SDGs—for example, decent work, reduced inequality, good governance and climate action—points to the mammoth task ahead, as well as the possibilities for driving internal and external improvements.

A growing number of universities are now working to embed the SDGs into their strategic plans, research activities, curriculum, pedagogy, student experiences, graduate attributes and institutional reporting. Some universities are focusing on a small subset of the SDGs, whilst others are taking them as a whole and considering their higher-order objectives. Here, we take the latter approach because no SDG can be ignored. The university sector's internal diversity is a unique match for the array of issues covered by the SDGs, and the agenda is designed to be integrative precisely because more reductionist approaches have helped generate the trouble we are in today.

Focused broadly on the agenda's two pillars—social justice and environmental sustainability—we explore how they intersect with the three paradoxes outlined earlier. Although a work in progress, our analysis to date has convinced us that universities have an important role to play in helping drive progressive and meaningful change, beginning with internalising them and applying them to their own operations and then reaching out to as many different groups as possible. There are many pathways and forms this could take.

Despite universities' many flaws and the deep legacy of development, including contemporary neoliberal notions of status and progress, we do not want to just jettison the idea and possibilities of the university for bringing about transformative change. Nor do we dismiss the potential of the SDGs, whose potentially transformative lines of flight are yet to be explored. However, to engage with the SDGs means bearing witness to the unsustainability of our current global conditions, the role of universities within this, and the discomfort, contradictions, tensions, fear, sadness, silences—as well as the creative spaces and transformative possibilities—that this can provoke.

Staying with the Trouble

The vision we outline in this book is of universities at the forefront of reflexive and critical thinking and action around the SDGs to both identify synergies and tensions and co-develop advice, activism and advocacy with a wide range of cross-sector stakeholders. We dare imagine this as the beginning of a progressive turn in higher education, one that uses the SDG agenda as a vehicle for transformative change. Underpinning this vision is an intellectual framework that approaches what universities and SDGs are and how they relate to each other and wider phenomenon as an open question, not an analytical starting point.

Our basic starting point is feminist and critical social science scholarship that not only identifies and tackles how to reframe and reshape fundamental problems in the world but also attends closely to the question of why, and in what ways, we should do so. Core to this is reflecting on the particular social configuration known as academia that we are part of and considering how it is interacting with—and could interact with— other aspects of the world. The goal here is to shift attention from a focus on the 'the what' to 'the how' of the SDGs as a university priority and agenda. We then go further to focus on the equally critical questions of 'why' engage with the SDGs and 'to what ends' does/will this serve progressive ends for the university and society.

Informed by pragmatic philosophy, this is about what Henrik Wagenaar calls attending to meaning-in-action.[41] It is about using a critical and dialogical lens that serves the common good. In other words, our focus is less on the universities and SDGs as abstract categories and more on what they are doing, or could do, in practice. Urban planning activist Leonie Sandercock describes this as a commitment to 'practising utopia' by taking a position on issues of democracy, power, social justice and sustainability within 'actually existing practices'. This in turn involves the development of a new dialectical imagination and the concomitant possibilities for both 'mobilizing hope' and 'negotiating fears' around a sustainable future.[42]

Many of the contemporary systems underpinning current unsustainability are robust, resilient assemblages (what Michel Foucault might call a dispositif),[43] held together in any one site or scale by a wide array of interlocking factors. From assumptions, norms and KPIs, to software programs, practices and rhythms, together these can make even the most critical and creative individuals feel like a cog in a machine. The challenge, therefore, should not be underestimated. At all levels, from individuals to universities to the planet, what is needed is better ways of surviving or coping. This goes far beyond how universities or people within them can 'be more resilient'—whatever that actually means.

Our collective ways of 'surviving' or 'coping' on this planet are far from sufficient in this climate of change. Stressors need to be neutralised, not normalised, and systems repaired and nurtured, not written off. Within universities such stressors extend far beyond the realm of the neoliberalism/s that many of us try to resist. Older conservative aspects of universities and global development agendas—including their classed, gendered and raced elements, close ties with the military and purported apoliticism—also need to be dismantled and replaced with inclusivity, reflexivity and transparency. Universities are also far from immune to other, more-than-economic global stressors, including the far-reaching effects of climate change, ocean acidification, biodiversity loss and pandemics.[44] Along with every other group of people, collection of places or set of practices, universities cannot function on a dysfunctional planet.

As Tristan McCowan notes, the university-society relationship is a complex one, 'involving the impact of the university on society (through the work and lives of its graduates, through the production of knowledge and through direct interaction with communities), but also the influence of society on the university, in a cyclical dynamic'.[45] When feedbacks onto and from the planet are added in, the relationship is especially complex. To address this reality, universities, along with every other organisation, need to not only 'do less harm' but 'do more good too'.[46] As well as neutralising stressors, all of us need to (re)generate positive futures.

Feminist scholar Donna Haraway cautions against turning away from the big challenges and argues we should instead 'stay with the trouble of living and dying in response-ability on a damaged earth'. This means bearing witness to the trouble of our times, rather than pinning our hopes on an imagined future that is decoupled from the monstrosity of the past and present.[47] To this end, her mobilisation and meaning of 'to trouble' is three-fold:

- Firstly, to recognise and accept that we live in troubling times, 'turbid, mixed up and disturbing';
- Secondly, that to change this we need to make trouble and 'stir up potent responses to devastating events'; and
- Finally, we need to then settle the troubled waters through the rebuilding of people, planet and place.[48]

As we have argued, the SDGs are not inherently static or repressive—unless we make them so—and nor are universities. There is scope to carve out the regenerative and transformative change we seek and need. Conceiving of universities and the SDGs as assemblages rather than stable, self-evident entities exposes the many aspects of each that remain beyond the reach of neoliberal efforts, or resilient to its impositions.

A similar stance is needed on neoliberalism and capitalism themselves, as scholars such as Sally Weller, J.K. Gibson-Graham and Brian Massumi have argued.[49] In contrast to disempowering images of either The Economy or The Market as all-encompassing and centrally controlled totalities, they are understood instead as messy assemblages that rely on being continually remade. Crucially, this means they are open to resistance, evolution and transformation, as efforts to recover from the COVID-19 disruption may demonstrate.

Universities are also diverse and messy, characterised by 'varieties of academic capitalism' among other things.[50] Similarly, universities could help change, contest and succeed neoliberalism and capitalism. Crucially, the SDG agenda offers a valuable tool in doing so. For example, the commitment to economic growth featured in SDG 8 (Decent Work and Economic Growth) is an opportunity to foreground and address the perverse effects of economic fundamentalism on other goals, the SDG agenda as a whole and the worlds we share.[51]

Universities need to engage much more deliberately with the real world they are part of, but real-world relevance is not about a hard-nosed, unthinking push to contribute more directly to economic growth. It is about acknowledging the complex material realities that universities have never left, and starting to reverse some of the damage universities and all of us within them have been complicit in generating. Following Simone de Beauvoir, 'It is in the knowledge of the genuine conditions of our lives that we must draw our strength to live and our reasons for acting'.[52]

Reworking the Matrix

The question then is how universities and the SDGs might be brought together to work in concert for positive transformational change. It is a question we begin to address in this book, acknowledging that wide and ongoing dialogue and experimentation is needed. We argue that the two crucial factors are: the depth and breadth of a university's and the sector's institutional commitment; and the ethics and boldness of their innovation culture, where innovation is understood as doing something differently and universities are understood as targets as well as sources of change. As outlined in Fig. 1.2, each can be mapped as an axis or continuum that yields four plausible scenarios that provide the basis for informed discussion about the strategic direction of university engagement with the SDGs: disengaged, paternalistic, tolerant or transformative.

The two axes of commitment and innovation represent two key uncertainties or questions: How deeply will a university commit to the SDGs? How bold and ethical will its innovation culture be? Where a university positions itself with regards to these two questions will determine its approach and potential for transformative change. Our aim is to provide some provocations to contribute to this dialogue, informed by our experiences of discussing and working with many others within universities about the SDGs, what they mean and how they might be used.

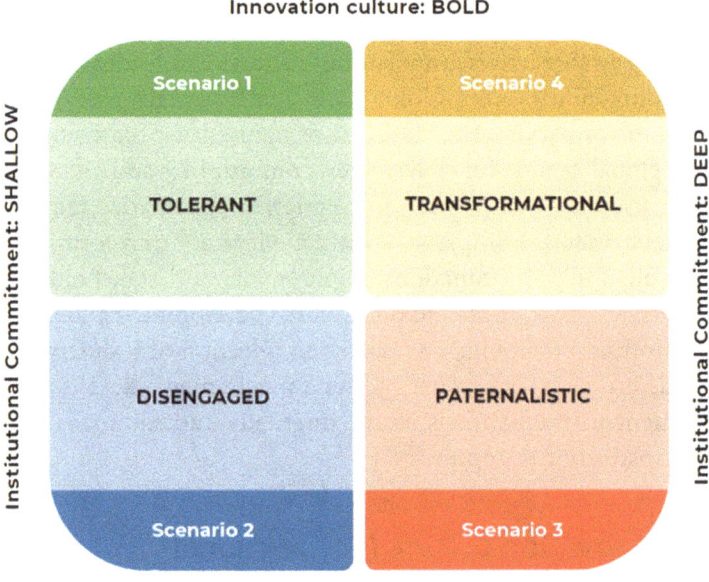

Fig. 1.2 Four possible scenarios for university engagement with the SDGs. (Source: Authors)

Axis 1: Institutional Commitment (Shallow to Deep)

At one end of the institutional commitment spectrum is Shallow Commitment which takes the form of tolerance for or occasional endorsement of SDG-related initiatives. Efforts around the SDGs may or may not exist in this scenario, but if they do, they are largely the work of isolated individuals or groups and are generally ad hoc, disconnected, invisible to most people, and quickly forgotten. They include one-off events, single assessment tasks or courses, occasional publications and short-lived research, operational projects or static webpages. The SDGs are treated (if at all) as a specialist topic and matter of personal interest, with limited relevance to the functioning of the institution's core business. More specifically, the SDGs are misunderstood by many people as simply a traditional international development issue and thus salient only to low-income countries and development specialists.

At the other end of the spectrum is Deep Commitment. Here, SDG engagement is characterised by strong institutional leadership, strategic prioritisation, cultural commitment and a critical pedagogy around progressive transformation. The SDGs are recognised and represented as part of a new global agenda for universities, communities and all professions. They are used as an integrative, long-term, systematic framework of engagement that encompasses—with a view to transforming where needed—all university functions, components and stakeholders. From the university's strategic plan to professional development and promotion of staff, from its resourcing of research to selection of industry partners, the SDGs are used as a cohering, focusing framework. The university commitment to the SDGs is visible internally and externally, with far-reaching institutional impacts.

Axis 2: Innovation Culture (Conventional to Ethical/Bold)

Universities may be deeply committed to SDG engagement across the institution, but still not do much differently, other than reshape their existing processes and practices. Cutting across the question of commitment is the question of an organisation's innovation culture, which can be characterised by how routine or imaginative it is. Routine innovation often involves the prolific production of innovation outputs developed in a conventional, competitive way, typically focused on technologies and business. One of the ironies of innovation is that as a concept it is far from novel. Indeed, it is now mainstream, often forced, and largely habitual, driven by an unthinking and seemingly inexorable need to produce new products (including academic papers) for the market. A conventional innovation culture perpetuates this robotic approach.

Situated at the other end of the innovation culture spectrum is a bolder, more radical approach to innovation that nurtures creative shifts and scales them out to generate uptake and to progressively alter, not reinforce, the existing institutional environment. Avoiding critique and change for their own sake, this approach involves innovating not just with products, but with ways of doing things, including innovation itself. The aim is to more explicitly, directly and effectively connect a university's work to meet society's needs.

Responsive to calls over the last two decades for a 'new social contract' for academia,[53] this attempted repositioning of universities involves a shift from top-down, linear, knowledge-centric models of innovation to more systemic, inclusive, action-oriented ways of doing innovation. It also responds to the growing realisation that the conventional approach to innovation is a source of problems as much as solutions, underlying many environmental harms and social injustices, as well as the unhelpful attitude in academia (mentioned above) of constant, competition-driven criticism of others.

An ethics-based approach to innovation is courageous, imaginative, generous and intelligent enough to not just change product specifications but also systems, goals and paradigms—including the innovation culture itself—so that societal needs and goals are more effectively met and people are nurtured along the way. Universities are being called upon to confront the effectiveness and ethics of their innovation strategies and practices. The challenge is to bring to the fore this ethical dimension and to confront it head on in order to better align activities with a desired ethical framework, such as the SDGs. The ethical innovation we propose can be summed up as *responsible, attentive, disruptive, authentic and regenerative*. We discuss ethical innovation in more detail in Chap. 4.

Future Scenarios

Though the future is unknown, it is highly likely that universities will be expected to more directly address the SDGs as issues such as climate change escalate. How, though, will any one institution respond? The two axes of commitment and innovation outlined above represent four possible scenarios, as shown above in Fig. 1.2. While these are clearly simplifications, each scenario provides a heuristic tool for thinking through options for a university and the implications of these choices.

Tolerant

The first scenario combines a shallow institutional commitment with a bold innovation culture. The tolerance pathway frames the SDGs as a specialist topic that some staff, students and partners are interested in, and are in fact doing creative and important things with. At the institutional level, the SDGs are resourced in a minor way, but are not recognised as a major societal challenge or guiding parameter, or as relevant to the institution as a whole. Instead, the university abides with some staff and students working in the area, reports diligently on the SDGs and cherry picks opportunities from the SDG agenda in keeping with its largely agnostic, opportunistic attitude to topic areas. Those actively working on the SDGs are largely left to their own devices, perhaps developing niches of radical innovation (e.g. bold experiments with partners in government, business and community), but in a generally isolated manner that is despite, not because of, what the rest of the university or academic sector is doing.

Disengaged

The second scenario represents a step backwards. It consists of a shallow institutional commitment and conventional innovation culture. Here, a university may commence work on the SDGs but it stagnates and fades over time, withering away to become just one of a number of reporting requirements and past enthusiasms. Some SDG work continues in the university, but it is largely ad hoc and driven by external requirements such as demands from funding bodies, industry partners and university ranking processes. Meanwhile, the university innovation culture is focused ever more narrowly on accelerating and refining existing product development processes and serving certain market players, while remaining disengaged from most of the society and the processes' wider ramifications. Individuals striving to do things differently are implicitly discouraged and will likely move on to other more open-minded institutions or sectors.

Paternalism

In the third scenario, a deeper institutional commitment is combined with a conventional innovation culture. The university works to embed the SDG agenda as a strategic priority from the top down across its four core functions of research, education, governance and operations, and external leadership. It takes the SDGs seriously as a moral obligation and/or as a pressure that the institution is compelled to adapt to, even if it is not convinced of the importance of revitalising sustainable development per se. As with the associated impact agenda, the university directs staff and students to engage with the SDGs in their work through a variety of compulsory and voluntary mechanisms including, for example, awareness raising, the inclusion of the SDGs as criteria in staff promotion processes, the resourcing of some SDG research initiatives and the incorporation of the SDG agenda into the institution's strategic plan. While some of these initiatives succeed in generating enthusiasm among some staff and students, others resist it as a bureaucratic imposition or adopt a minimal compliance mindset.

Transformational

The final scenario—the one we want to help generate through this book and other efforts—is focused on the need for transformation. It combines a deep institutional commitment and a bold, ethical innovation culture.

The transformational scenario involves a critical, ethical commitment to rapidly transitioning the university into a better position in order to help transition the world onto a more sustainable, socially just pathway. It commits to the principles and ethos of ethical innovation and works determinedly to scale bold, ethical innovations for sustainable development up and out, both across the University—from domain to domain, project to project, process to process, course to course—and across its stakeholder places, organisations and sectors. This institutional commitment is deep, bold and pioneering, showcasing and sharing different epistemological understandings and pedagogical practices, underpinned by visionary leadership, resources and support. If not now, then when?

Pathways and Provocations

The chapters of this book call for transformational change for universities in a world in crisis. The pathways and provocations of the book position the SDGs as a critical, regenerative lens for universities and higher education: an orientation and orienting device—outwards and inwards—to the past, and to a more positive future. The emphasis is not only about what universities can do for SDGs (although this is clearly important) but also about what SDGs can do for universities given their shared 'developmentality', neoliberal legacies and boundary-crossing character.

The first half of the book lays out the intellectual framework and practical agenda driving the book. This chapter outlines our critical approach to the SDGs as a witness statement to the unsustainability of modern development (including in higher education). Our starting point is feminist and critical social science scholarship that seeks to reframe and reshape the dominant developmentalities but also attends closely to the question of why and in what ways we should do so. The goal here is to shift attention from a focus on the 'the what' to 'the how' of the SDGs as a university priority and agenda. We then go further to focus on the equally critical questions of 'why' engage with the SDGs and 'to what ends' does/will this serve progressive ends for the university and society.

Core to this is reflecting on the particular social configuration known as academia that we are part of and considering how it is interacting with—and could interact with—other aspects of the world. We make the case for universities to embrace a deep commitment to the SDGs combined with a bold, ethical innovation culture. This would lead to transformational change in and through organisations and the academic sector if operationalised effectively. It represents the scaling up of an idea—such as the SDGs—from a niche concept into the workings of institutions, up through the levels of governance that scales deep and wide. The SDGs become embedded in everything universities do—as critical co-production and regenerative assemblages.

In Chap. 2 we turn more explicitly to the evolving role of the SDGs within the context of the Anthropocene. The story of the SDG agenda is a story about development and the relationship between the present and the future. Not only does the SDG agenda aim to shift existing development trajectories but the way it is itself narrated by groups such as the UNDP above (the United Nations Development Program) casts it as a positive development in and of itself, as a kind of awakening and new age. What the SDGs do in practice, however, is far from certain or predetermined. Shaping its actual outcomes are legacies from the past, competing worldviews and different readings of the sustainable development challenge.

Within the context of the Anthropocene, the 2030 SDG agenda represents the goal posts we jointly need to orient towards and to find ways of working differently. These goal posts are wide and diverse but represent a significant shift from the *status quo* within both universities and society. Encompassing action on climate change, transformational innovation, resilient infrastructure, economic progress, gender equity, good governance and environmental sustainability, the SDGs represent a new standard by which good practice and success are now being understood and measured. They are stimulating interest in alternatives to dominant modes of development (including those within the university). The Indigenous Latin American worldview *Buen vivir* (living well) for example resonates with aspects of the SDGs[54] and invites ways to re-imagine higher education that pushes beyond the limitations of the neoliberal ideology committed to economic growth at the expense of the environment.

Chapter 3 directs and develops the focus on the reciprocal role of the universities and the SDGs as both process and outcome (i.e. means *and* ends) in the age of disruption, crisis and change. Moving beyond the nationalistic and individualistic competitor mindset, the SDGs encourage universities to heed the global call to action. Universities are vital to progressing the SDG agenda—both as large organisations in their own right and as enablers of others. They have a fundamental role to play across all four of their functions: learning and teaching, research impact, external leadership and internal operations. In the twenty-first century, universities have the opportunity and capacity to move into a leading position in supporting and promoting sustainability through research, education, external leadership and governance. This goes beyond mapping existing SDG capabilities, to embedding sustainability vertically and horizontally across diverse communities of practice, sectors and scales.

The main argument of Chap. 3 is that universities are not isolated ivory towers, floating free from the rest of the world. As their remaking as corporations over the last fifty years illustrates, they are 'of the world'. For better or worse, they are being constantly reshaped by the world and, for better or worse, they are continually shaping the world—in ways that far exceed laborious efforts at 'engagement'. Universities are not just enablers of change in the SDG agenda but also important targets of change. Whether conceived as primarily members of the public or private sector, universities are large organisations/institutions with a wide range of internal functions and responsibilities with far-reaching implications for the SDG agenda. For universities to perform their unique function as enablers of change, they need to simultaneously embrace their role as targets for change and ensure they are role modelling the sort of approaches and impacts they want to engender.

Chapter 4 outlines and articulates the principles underpinning 'Ethical Innovation' as a normative frame for higher education, that is, Responsible, Authentic, Disruptive, Adaptive, Regenerative (RADAR). The urgency and complexity of sustainable development means universities need to be more energetic *and* careful in generating change. There is a growing realisation that universities need to start taking questions about their purpose and approach more seriously. In this chapter we build on this by looking in more detail at the question of how universities might work in a way that is resonant with the transformative aspirations of the SDGs. The aim is not to provide a blueprint for how universities can engage with the SDG agenda specifically but to move to the next question of how universities and those within them can *create the enabling conditions* needed to orient towards the SDGs.

To do so we look at complementary strategies for generating these enabling conditions with a focus on cultivating ethical innovation encompassing all areas of university activity. Ideas and practices around innovation and impact are intimately related to the base concept of development, and both point to the need to reclaim the concept, calling out contemporary conventional development as capitalist development and introducing the sort of *regenerative* development that the world badly needs, including universities. By inventing or legitimating some realities and not others, and being shaped in turn by those, knowledge production helps co-produce the world. With the world now in an increasingly de-generate state, there is an overdue need to critically evaluate this power and responsibility. In particular there is a need to examine how knowledge production and dissemination within universities has helped generate and is continuing to generate the current world from micro to macro scales, and to explore how it could re-generate more habitable and humane ones.

The second half of the book focuses on how the SDGs and higher education are co-produced in practice and the prospects for transformative change. Chapter 5 emphasises the role of research as an evolving development ethos and double-edged sword. Existing dominant approaches to university research are not adequately meeting societal and planetary needs as outlined in the SDGs. Nor are they meeting societal expectations or building public trust. Wider community expectations of what higher education can and should be within society are shifting. This includes growing calls to re-imagine what success looks like for higher education in the quest for the 'good university' driven by social good rather than profit to build sustainable societies.

Research is development-like, but—by positioning itself as a purported distant observer or disguising itself as a mere processor of others' values and wishes—it has not been subject to the sort of fierce reflexivity and renovations that social and economic development have. As a development process, research now urgently needs to become more *like* sustainable development if it is to contribute usefully to sustainable development. Regardless of the topic area, discipline or institution, research needs to become more aware of complexity, uncertainty and the deeply political and ethical nature of all research endeavours (including those endeavours that are conspicuous in their absence), as well as its concomitant role in a sustainable future.

In Chap. 6 the significance and importance of learning and teaching (L&T) as critical pedagogy *about, for* and *through* the SDGs is explored. Understandings and practices around L&T are evolving to better address the need for meaningful real-world change. As educators this is an opportunity to attune to what is most important and to do what we do best. It is about pausing to ask hard questions about what the world needs and not simply what the market wants now. It is about celebrating what educators in universities are able to contribute by leveraging the power of our deep knowledge, academic networks and independence to not only do practical applied research of the sort many research actors can do but identify neglected issues and voices, articulate lessons from the past, critique existing approaches and anticipate possible futures including the shift to on-line modes of engagement.

There is a need to critically engage with what 'transformational' education means in the context of universities and their reciprocal engagement with the SDGs. Embedded with a critical praxis and building on the work of Education for Sustainable Development (ESD), these transformational approaches are likely to be those that are student-driven, interdisciplinary and boundary crossing, with a strong emphasis on participatory approaches to knowledge, co-creation, generation and acquisition. L&T within the context of the SDGs is not value free, but a critical, ethical agenda focused on the changes required for a more sustainable future. The emphasis following the critical pedagogy of Paulo Freire is not just on 'what' is the L&T content, but 'how' and 'why' L&T in higher education matters, in what ways and for whom within the context of a rapidly heating and increasingly inequitable planet.

Chapter 7 is based around two questions that are often raised within the context of university engagement with the SDGs: What does success look like? How would we know? To respond to this the chapter takes up the provocation of 'The Good University' and revisits the matrix and scenarios for transformative change around the SDGs in higher education. The idea of the modern university is a contested vision in a climate of growth-led change. The impacts of marketisation, globalisation and massification have created unprecedented shifts in both the real and perceived contributions of the university including the increasingly contested role of metrics data and indicators as measures of success.

Wider community expectations of what higher education can and should be within society are shifting. This includes growing calls to re-imagine what success looks like in the quest for the 'good university' driven by social good rather than profit, to build sustainable and just societies that are able to co-exist within a healthy planet. What constitutes success and impact is constantly evolving—and will continue to do so—as a result of the shifting relationships between universities and society. Partnerships and processes are complex and relational and premised on the need for ethical innovation and commitment to achieve the transformative ambitions of the SDGs. Critical understandings and practices of what success looks like as a reciprocal agenda for universities in relation to advancing the SDGs must be articulated and are necessarily contested and mutually shaping. 'Becoming sustainable' must evolve in ways that better address meaningful real-world change.

The final Chap. 8 summarises ways to build capacity and momentum around the SDGs across the university—intellectually, practically and culturally. There is a substantial gap between academic-based, real-world-engaged approaches that catalyse positive action across sectors and business as usual. Addressing this involves strategies to harness the vital work already underway in higher education institutions, as well as frameworks for fostering new initiatives to trigger and scale up ethical innovation across the university. Whatever their size, shape, scale or funding model, or their capacity to cultivate and share ideas, methods and frameworks for the betterment of society—universities matter as formalised 'critical space' and agents of change.

Universities are committed to a public mission that underpins their purpose and function in society: as centres of new knowledge, understanding, skills and experience, through research, learning and teaching, leadership, outreach and service to society. As proponents of progress, choice, debate and engagement, universities can set the pace for the SDGs in the key areas of society, culture, economics and the environment. They are strategic incubators for policy, research and advocacy, education and training, and professional and community engagement. In building a transformative agenda around the SDGs, higher education works to nurture niche initiatives that build on, link and extend existing work and build individuals' agency, as part of the critical changes needed to embed the transformative ethos of sustainability into the university structures and development processes.

Addressing *sustainable* development in the Anthropocene is not about tinkering around the edges. Just as development cannot be fixed with international development add-ons, sustainability cannot be addressed with green add-ons. Shallow or tokenistic engagement with the SDGs risks distracting from and legitimating business as usual, thereby perpetuating the unsustainabilities that are pushing us towards deeper injustice and planetary collapse. Universities are as guilty of cynical, inauthentic engagement with the SDGs as any other institution. But they are also animated by an inherent future-focus, one that is core to their developmentality. The radical uncertainties of the Anthropocene do nothing to dim this focus, but they do blur our vision and demand we also look backwards, all around and into our institutions and selves to understand the situation we are in—and question what it is we are trying to develop. This is the transformative SDG agenda we imagine: a critical, ethical regenerative politics and praxis that seeks to reshape dominant development trajectories including those within higher education. A witness statement that constantly reminds us that other more sustainable futures are still possible.

Notes

1. Attenborough, D (2020) *A Life on our Planet*, Netflix Film, available on https://www.netflix.com/au/title/80216393
2. Beauchamp, C (2020) *David Attenborough's Witness Statement: 'A Life on our Planet'*, 13th October, accessed online at https://impakter.com/david-attenboroughs-witness-statement-a-life-on-our-planet/

3. Salazar, J.F. (2015) Buen Vivir: South America's rethinking of the future we want, *The Conversation*, 24th July, accessed on https://theconversation.com/buen-vivir-south-americas-rethinking-of-the-future-we-want-44507
4. Uluru Statement of the Heart (2017) accessed online at https://fromtheheart.com.au/uluru-statement/the-statement/
5. Kipnis, L (2003) *Against Love*, New York, Random House.
6. van Loon, J (2019) *The Thinking Woman*, Sydney, NewSouth Publishing.
7. Lorde, A (1984). The Master's Tools Will Never Dismantle the Master's House. *Sister Outsider: Essays and Speeches*. Berkeley, CA: Crossing Press. p. 110–114.
8. A challenge for geographers eloquently outlined by Diana Liverman (2018) in Geographic perspectives on development goals: Constructive engagements and critical perspectives on the MDGs and the SDGs, *Dialogues in Human Geography*, 8(2), p. 168–185.
9. Hajer, M., M. Nilsson, K. Raworth, P. Bakker, F. Berkhout, Y. de Boer, J. Rockström, K. Ludwig and M. Kok (2015) Beyond cockpit-ism: Four insights to enhance the transformative potential of the sustainable development goals. Sustainability 7(2), p. 1651–1660.
10. Zelizer, B (2002) Finding aids to the past: bearing personal witness to traumatic public events, *Media, Culture & Society*, 24, p. 697–714.
11. Felman, S. (1992) The Return of the Voice: Claude Lanzmann's Shoah, in S. Felman and D. Laub (Eds.) (1992) *Testimony: Crises of Witnessing in Literature, Psychoanalysis and History*. New York: Routledge, p. 204.
12. Deleuze, G., and Guattari, F. (1987). *A thousand plateaus: capitalism and schizophrenia*. Minneapolis: Minnesota Press.
13. Braidotti, R (2013) *The Posthuman*, Cambridge, Polity Press, p. 104.
14. Moore, M.-L., Riddell, D., Vocisano, D. (2015) Scaling out, scaling up, scaling deep: strategies of non-profits in advancing systemic social innovation. Journal of Corporate Citizenship, 67–84.
15. Kamola, I (2016) Situating the "global university" in South Africa, in Chou, M, Kamola, I and Pietsch, T (eds.) *The Transnational politics of higher education: Contesting the global/transforming the local*, Routledge, New York, p. 42–63.
16. Steele, W (2020) *Planning Wild Cities: Human-nature relationships in the urban age*, London/New York, Routledge.
17. Pengilley, V (2018) From vampires to zombies- the monsters we create say a lot about us, ABC Radio National, 9th September, accessed on https://www.abc.net.au/news/2018-09-09/monsters-we-create-reflect-our-fears-and-desires/10174880

18. Finn, E (2013) *The Corporate Frankenstein Monster*, in Canadian Centre for Policy Alternatives, accessed on https://www.policyalternatives.ca/publications/monitor/corporate-frankenstein-monster
19. Halovitch, H (2019) 'Vampires and Ratko Mladić: Balkan Monsters and The Monstering of People', in Lee, J, Halilovitch, H, Landau-Ward, A, Phipps, P and Sutcliffe, R (2019) *Monsters of Modernity: Global icons for our critical condition*, Kismet Press.
20. Kaika, M (2017) "Don't call me resilient again!" The new urban order as immunology...or what happens when communities refuse to be vaccinated with 'smart cities' and indicators, *Environment and Urbanization*, 20 (1), p. 89–102.
21. Kopnina (2020) Education for the future? Critical evaluation of education for the sustainable development goals, *The Journal of Environmental Education*, 51(4), p. 280–291.
22. Slocum, S, Diitrov, D, Webb, K (2019) The impact of neoliberalism on higher education tourism programs: Meeting the 2030 sustainable development goals with the next generation, *Tourism management perspectives*, 30, p. 33–42.
23. Blythe, J, Silver, J, Evans, L, Armitage, D, Bennett, N, Moore, M, Morrison, T, Brown, K (2018) The darkside of transformation: Latent risks in contemporary sustainability discourse, *Antipode*, p. 1–14.
24. Pengilley, V (2018) From vampires to zombies—the monsters we create say a lot about us, ABC Radio National, 9th September, accessed on https://www.abc.net.au/news/2018-09-09/monsters-we-create-reflect-our-fears-and-desires/10174880
25. Hope, J. (2021) The anti-politics of sustainable development: Environmental critique from assemblage thinking in Bolivia. *Transactions of the Institute of British Geographers*. Online first.
26. Carrol, N (2002) Why Horror? in Jancovich, N (Ed), *Horror: The Film Reader*, Routledge, New York.
27. Rickards, L. (2020) Ironies of the Anthropocene. In: D. Chandler, K. Grove and S. Wakefield. *Resilience in the Anthropocene: Governance and Politics at the End of the World*. London, Taylor and Francis.
28. Morton, T. (2007) *Ecology Without Nature: Rethinking Environmental Aesthetics*. Cambridge, MA: Harvard Univ Press. p. 21.
29. Gibbs, R. W. (2002) Irony in the wake of tragedy. *Metaphor and Symbol* 17: 145–153. p. 152.

30. Szerszynski, B. (2007) The post-ecologist condition: irony as symptom and cure. *Environmental Politics* 16: 337–355. p. 351.
31. Blühdorn, I. (2011) The politics of unsustainability: COP15, post-ecologism, and the ecological paradox. *Organization & Environment* 24: 34–53.
32. Barry, J. (2013). *The Politics of Actually Existing Unsustainability: Human flourishing in a climate-changed, carbon-constrained world.* Oxford, Oxford University Press.
33. In part this is due to the financialisation of the university sector and normalisation of debt. See Eaton, C., Habinek, J., Goldstein, A., Dioun, C., Santibáñez Godoy, D.G., Osley-Thomas, R. (2016) The financialization of US higher education. Socio-Economic Review 14, 507–535. Notably, such financialisation parallels movesinancialise SDG delivery. See Mawdsley, E., (2021) Development Finance and the 2030 Goals, in: Chaturvedi, S., Janus, H., Klingebiel, S., Li, X., Mello e Souza, A.d., Sidiropoulos, E., Wehrmann, D. (Eds.), The Palgrave Handbook of Development Cooperation for Achieving the 2030 Agenda: Contested Collaboration. Springer International Publishing, Cham, pp. 51–57.
34. Rogoff, I. (2003) From Criticism to Critique to Criticality. *Transversal Texts* https://transversal.at/transversal/0806/rogoff1/en. p. 1.
35. Roseneil, S. (2011) Criticality, Not Paranoia: A Generative Register for Feminist Social Research. *NORA—Nordic Journal of Feminist and Gender Research* 19, 124–131.
36. Rogoff (2003) p. 1.
37. Moseley, W.G. (2018) Geography and engagement with UN development goals: Rethinking development or perpetuating the status quo? *Dialogues in Human Geography,* 8(2), p. 201–205.
38. Meyer, R. (1968) Language: Truth and Illusion in "Who's Afraid of Virginia Woolf?". *Educational Theatre Journal* 20, p. 60–69.
39. Billington, M (2016) Who's Afraid of Virginia Wolf? is a misunderstood piece, *The Guardian,* 18th September accessed on https://www.theguardian.com/stage/2016/sep/18/whos-afraid-of-virginia-woolf-edward-albee
40. Sultana, F. (2018) An(Other) geographical critique of development and SDGs. Dialogues in Human Geography 8 (2), p. 186–190.
41. Waagner, H (2011) *Meaning in Action: Interpretation and dialogue in policy analysis,* New York, M.E Sharpe Inc.

42. Sandercock, L (2002) Practicing utopia: Sustaining cities, *Disp-P The Planning Review* 38(148), p. 4–9.
43. See Foucault, M (1977) *Discipline and Punishment: The birth of the prison,* New York, Pantheon Books.
44. Rickards, L., Watson, J.E. (2020) Research is not immune to climate change. *Nature Climate Change* 10, 180–183.
45. McCowan, T. (2019). *Higher Education for and Beyond the Sustainable Development Goals,* Springer. p. 17.
46. Buhmann, K., Jonsson, J., Fisker, M. (2019) Do no harm and do more good too: Connecting the SDGs with business and human rights and political CSR theory. *Corporate Governance: The International Journal of Business in Society,* 19(3), p. 389–403.
47. Haraway, D (2016), *Staying with the trouble: Making kin in the Chthulucene,* Duke University Press, Durham.
48. Ibid.
49. See Weller, S and O'Neill, P (2014) An argument with neoliberalism: Australia's place in a global imaginary, *Dialogues in Human Geography,* 4(2), 105–130. Gibson-Graham, J.K. (2006) *A Post-capitalist politics,* Minnesota, Uni of Minnesota Press, Minneapolis. Massumi, B. (2018) 99 Theses on the revaluation of value: A postcapitalist manifesto. Uni of Minnesota Press, Minneapolis.
50. Schulze-Cleven, T., Olson, J.R. (2017) Worlds of higher education transformed: toward varieties of academic capitalism. *Higher Education* 73, p. 813–831.
51. Boldeman, L. (2007) *The Cult of the Market: Economic fundamentalism and its discontents.* ANU Press, Canberra, Australia.
52. de Beauvoir, S (2015) *The Second Sex,* London, Penguin.
53. See for example Lubchenco, J. (1998). "Entering the Century of the Environment: A New Social Contract for Science." *Science* 279(5350), p. 491–497.
54. Death, C., Gabay, C. (2015) Doing Biopolitics Differently? Radical Potential in the Post-2015 MDG and SDG Debates. *Globalizations* 12, p. 597–612.

2

Sustainable Development in the Anthropocene

The Story of the SDGs

In the year 2015, leaders from 193 countries of the world came together to face the future. And what they saw was daunting. Famines. Drought. Wars. Plagues. Poverty. Not just in some faraway place but in their own cities and towns and villages. They knew things didn't have to be this way. They knew we had enough food to feed the world, but that it wasn't getting shared. They knew there were medicines for HIV and other diseases, but they cost a lot. They knew that earthquakes and floods were inevitable, but that the high death tolls were not. They also knew that billions of people worldwide shared their hope for a better future. So leaders from these countries created a plan called the Sustainable Development Goals (SDGs). This set of 17 goals imagines a future just 15 years off that would be rid of poverty and hunger, and safe from the worst effects of climate change. It's an ambitious plan.[1]

The story of the SDG agenda is a story about development, which is to say it is a story about the relationship between the past, present and the future. Not only does the SDG agenda aim to shift existing development trajectories, but the way it is itself narrated by groups such as the UNDP above (the United Nations Development Program) casts it as a positive development in and of itself, as a kind of awakening and new age. What the agenda does in practice, however, is far from certain or predetermined. Shaping its actual outcomes are legacies from the past, competing worldviews and different readings of the sustainable development challenge.

This is not the first time that the world has had a set of global goals. Immediately preceding the SDGs were the eight Millennium Development Goals (MDGs), established at the turn of the Millennium to much fanfare. As a final report on the MDGs describes: 'At the beginning of the new millennium, world leaders gathered at the United Nations to shape a broad vision to fight poverty in its many dimensions. That vision, which was translated into eight Millennium Development Goals (MDGs), has remained the overarching development framework for the world for the past 15 years.'[2]

As the epitaph above demonstrates, the strongly normative discourse about *shared* problems and heroic action that shaped the MDGs is continued with the SDGs. Thematically the SDGs also build on the MDGs, incorporating the issues highlighted by the MDGs within the new framework in recognition of the enormous amount of work still needed to properly address the problems they name (see Table 2.1).

Despite the similarities and overlaps, the SDG agenda differs in three main ways.

First, the SDGs substantially broaden the range of issues included, expanding the number of goals from eight to seventeen. While three health-related MDGs are rolled into SDG 3 Good Health and Wellbeing, some are disaggregated, such as MDG 1 Eradicate extreme poverty and hunger, which is broken into the first two SDGs, and MDG 7 Ensure environmental sustainability, which is distributed across multiple SDGs, including SDG 6 on Clean Water and Sanitation and SDG 13 on Climate Action. In addition, numerous other goals and ambitions are added to make explicit the need to tackle critical 'background issues' such as access to energy and post-primary education, unjust work conditions, and violence and conflict. For instance, SDG 9 Reducing Inequalities, plus broader attention to inequalities across the SDGs, explicitly recognises the fact that inequality in income, wealth and access to environmental goods and services between and within countries is persistent, even

Table 2.1 The 17 SDGs and the 8 MDGs

Sustainable Development Goal (2015–2030)	Description	Related Millennium Development Goal (2000–2015)
1. No poverty	End extreme poverty in all forms by 2030	MDG 1. Eradicate extreme poverty and hunger
2. Zero hunger	End hunger, achieve food security and improved nutrition, and promote sustainable agriculture	MDG 1. Eradicate extreme poverty and hunger
3. Good health and wellbeing	Ensure healthy lives and promote wellbeing for all at all ages	MDG 4. Reduce child mortality MDG 5. Improve maternal health MDG 6. Combat HIV/AIDS, malaria and other diseases
4. Quality education	Ensure inclusive and equitable quality education and promote lifelong learning opportunities for all	MDG 2. Achieve universal primary education
5. Gender equality	Achieve gender equality and empower all women and girls	MDG 3. Promote gender equality and empower women
6. Clean water and sanitation	Ensure availability and sustainable management of water and sanitation for all	MDG 7. Ensure environmental sustainability
7. Affordable and clean energy	Ensure access to affordable, reliable, sustainable and modern energy for all	
8. Decent work and economic growth	Promote sustained, inclusive and sustainable economic growth, full and productive employment and decent work for all	
9. Industry, innovation and infrastructure	Build resilient infrastructure, promote inclusive and sustainable industrialisation and foster innovation	
10. Reduced inequalities	Reduce inequalities within and among countries	
11. Sustainable cities and communities	Make cities and human settlements inclusive, safe, resilient and sustainable	MDG 7. Ensure environmental sustainability

(continued)

Table 2.1 (continued)

Sustainable Development Goal (2015–2030)	Description	Related Millennium Development Goal (2000–2015)
12. Responsible consumption and production	Ensure sustainable consumption and production patterns	MDG 7. Ensure environmental sustainability
13. Climate action	Take urgent action to combat climate change and its impacts	MDG 7. Ensure environmental sustainability
1. Life below water	Conserve and sustainably use the oceans, seas and marine resources for sustainable development	MDG 7. Ensure environmental sustainability
2. Life on land	Protect, restore and promote sustainable use of terrestrial ecosystems, sustainably manage forests, combat desertification, and halt and reverse land degradation, and halt biodiversity loss	MDG 7. Ensure environmental sustainability
3. Peace, justice and strong institutions	Promote peaceful and inclusive societies for sustainable development, provide access to justice for all and build effective, accountable and inclusive institutions at all levels	
4. Partnerships for the goals	Strengthen the means of implementation and revitalise the global partnership for sustainable development	MDG 8. Develop a global partnership for development

worsening, and is a major inhibitor of good development outcomes.[3] Overall, the SDGs offer a far more comprehensive set of goals than the MDGs. As discussed below, this move to cover more (if not all) bases resonates with both the contemporary rise of systems thinking and an older development ideal.

Second, more radically, the SDGs do not just slice, dice and extend the list of issues covered, they reframe the development challenge more holistically, reflecting the paradigm of sustainable development that tries to integrate environment, society and economy. In so doing, they add not just a more systematic but a systemic approach, bringing into view the interconnections between processes in different areas, populations and sectors. Unusually, the SDG agenda attempts to tackle at least some causal drivers of contemporary problems (e.g. unsustainable consumption and production, unsustainable food systems, dirty energy sources) not just 'symptoms' such as environmental degradation, climate change and hunger. Its openness to facing some of the hard facts about contemporary society is one reason we see the SDGs as an opportunity for transformational change, especially if those in higher education and others can help push conversation and action further towards deeper root causes such as capitalism and colonialism.

Partly as a result of being presented as indivisible, the SDGs are also less spatially contained to certain regions. The SDG agenda is promoted as applicable to *all* groups everywhere, both in terms of where action is needed and who needs to be involved. As described below, no longer is the underlying model of development simply that of *international development* (the rich helping the poor 'catch up'), although strong elements of this approach do remain. It is also *global sustainable development*, where problems are seen everywhere, including problems generated by the rich, such as resource consumption and production practices that contribute significantly to serious negative social, economic and environmental 'externalities' in low-income areas.[4] Although, as critics have pointed out, opportunities to really mark wealthy populations and Western lifestyles as problematic were side-stepped in the agenda (e.g. malnutrition targets only include under-nutrition, not over-nutrition), the agenda is unusually overt in problematising elements of the conventional progress ideal, which is one reason the SDGs hold such far-reaching implications for universities.

Third, the SDGs reverberate with the urgent tone and planetary focus of recent intellectual and policy developments, notably discussion of the Anthropocene, planetary boundaries and resilience, and other major international agreements such as the Paris Climate Accord. Rather than the SDGs 2030 deadline simply being an automatic administrative reset of the 15-year period of the MDGs, 2030 is given real meaning in the SDG agenda due to growing awareness that the world is running out of time to avert runaway climate change and Earth System collapse. Like the Paris Climate Agreement, which it explicitly cross-references, the SDG agenda also began in 2015 and is similarly monitored in terms of likely outcomes in 2030. Failure to substantially reduce greenhouse gas emissions by 2030 will lock in dangerous levels of climate change, pushing the world beyond the target of 1.5 °C of average global warming and undermining the entire SDG agenda.

The 2020 UNEP Emissions Gap report on countries' voluntary commitments to mitigate greenhouse gas emissions points out that compared to what is needed to limit temperature rise to 1.5 °C, as of mid-2020 policy commitments across the world:

> remain seriously inadequate to achieve the climate goals of the Paris Agreement and would lead to a temperature increase of at least 3 °C by the end of the century.[5]

The report concludes that while more and more countries are committing to reducing emissions from their own activities to net zero, this leaves many emissions untroubled or unaccounted for and 'a dramatic strengthening of ambition is needed'. Specifically, countries collectively need to commit to five times the existing level of mitigation effort if we are to keep global warming to 1.5 °C.[6] Furthermore, distant policy commitments need to be translated into action now.

Despite a small dip in emissions due to COVID-19, actual emissions are still far in excess of even existing inadequate policy commitments, rising in 2019 to unprecedented levels, partly because of emissions from the growing number of forest fires that climate change feedbacks are exacerbating.[7] Global average temperature is already more than 1.15 °C above the pre-industrial average (1800–2019)[8] and reached a record high (equal with 2016) in 2020.[9] Combined with the way that far-reaching climatic changes and their cascading impacts are already eroding societal wellbeing, ecosystem health and institutional capacity, the situation is increasingly urgent. Many scientists are arguing more and more forcefully that every year—even every month—needs to achieve substantial greenhouse gas mitigation.[10]

Failure to achieve the SDGs will severely undermine society's capacity to mitigate future climate change rapidly and effectively enough. It will also undermine our capacity to cope with and adapt to the attendant climatic changes and pervasive flow-on effects.[11] How we and our communities, workplaces, institutions, landscapes and other living things are impacted by climate change is as much a matter of the 'conditions on the ground' that we are facing at a given moment in time as it is by climatic factors.[12] Such conditions are, in turn, an expression of not only prior specific climate adaptation actions (e.g. urban greening to reduce heat and flood risk, improved emergency communication systems) but the degree to which the myriad dimensions of sustainable development have been achieved in a given context, or not. Sustainable development is vital to successfully managing as well as avoiding climate change, and climate change action is an enabler and beneficiary of all of the SDGs, not just the focus of a single SDG (SDG 13).

Action on other SDGs is no less urgent than that on SDG 13, and not only because many of them—such as SDG 11 on sustainable cities, SDG 7 on clean energy, SDG 9 on responsible consumption and production and SDG 2 on sustainable food systems—are vital to lowering atmospheric greenhouse gas concentrations and/or vulnerability to climate change impacts. For example, biodiversity loss, which is explicitly covered in SDG 14 Life on Land and SDG 15 Life Under Water, is now so dire that it constitutes what some have declared a Sixth Mass Extinction in Earth's history. The 2019 Global Assessment Report by the Intergovernmental Panel on Biodiversity and Ecosystem Services—the first of its kind in nearly 15 years—concludes that despite overwhelming evidence that non-human nature is foundational to human wellbeing, 'the great majority of indicators of ecosystems and biodiversity' show 'rapid decline' since 1970.[13]

Underlying this reduction in the quantity and quality of biodiversity is the fact that pollution and invasive alien species are increasing, species assemblages are becoming more homogenised, and 'human actions threaten more species with global extinction now than ever before'.[14] The consequences are not limited to the non-humans involved in SDG 14. Rather, because 'Nature is essential for human existence and good quality of life', it 'is essential for achieving the Sustainable Development Goals'. The loss of ecological services such as clean air and water, temperature control, pollination, food and pharmaceuticals profoundly undermine the SDGs' progress. Conversely, progress on the SDGs is essential to conservation of nature and ecological services, demonstrating once again their reciprocal character.[15]

COVID-19 and its far-reaching flow-on effects are further highlighting the urgency and challenges of the SDG agenda. Among other things, the pandemic catastrophe is shining a harsh light on current human-animal interactions, global supply chains and spatial and social inequalities in health, health care, employment, social services, governance systems and green space. Lack of progress on the SDGs has exacerbated and co-generated the effects of the pandemic, while 'COVID-19 will likely negatively impact progress towards most SDGs in the short and medium-term, including in high-income countries'.[16] Despite or partly because of the disruption of the pandemic, the 'turn to the future' that the SDG agenda encourages is only strengthening. As we discuss below, this includes experimentation with different modes of imagining and governing the future.

Reactions and approaches to the SDG agenda vary widely, reflecting underlying worldviews, concerns and interpretations of component ideas. To help explain some of the key arguments, we now turn to the past to revisit the agenda's underpinning ideas. We begin with the very notion of development itself which has been interpreted and approached in radically different ways. The role of interpretation and implementation means that, like universities, the SDG agenda is not fully determined. Thus, its potential cannot be dismissed or bounded from the outset and is up to us to realise. At the same time, it is important to be aware of the baggage that development and related concepts carry. This means reflecting on questions of progress and sustainability, and their roots in big ideas and drivers such as modernity and colonialism.[17] In this chapter we look at international development, post-development and sustainable development in the Age of the Anthropocene, before sketching out some of the implications for universities.

The Concept of Development

At the heart of both the Sustainable *Development* Goals and the higher education sector is the idea of development. For the last few decades, the ambition and practices of development have been contested and regularly declared outdated, reflecting its long history. Despite claims by some that it is 'dead', development remains a highly resilient concept, in part because of its reincarnation as sustainable development and, more recently, its ongoing reworking in contemporary international discourse, such as the idea of 'climate-resilient development'.[18] Appreciating how development arose and functions as a concept is a crucial first step in critically engaging with the SDGs and understanding the current juncture.

Since at least the colonial period, the concept of development has become a basic pillar of Western thought and global governance, one with the 'power to frame our thinking of what is desirable and doable, and how'.[19] Being Western in origin, the concept of development and its associated measures and metrics have been used repeatedly to arrange the world's regions into an imagined temporal sequence in which Western, usually wealthy nations are designed as 'developed' (advanced) and others are more or less relatively 'less developed'.[20] It is this imagined temporal unevenness between (and to a lesser extent within) nations that has classically animated development initiatives and informs one of the cornerstones of the SDG agenda: 'leave no one behind'.

Although development has multiple historical roots and context-specific interpretations and uses, Finnish development studies scholar Juhani Koponen argues that the concept is characterised by three overarching and mutually reinforcing meanings: (1) 'a desired goal, an ideal state of affairs to strive for'; (2) 'a transformative process or, rather, a set of processes towards that goal'; and (3) 'intentional human action based on the belief that a well-meant intervention will trigger processes leading to what we ideally regard as development'.[21] Underpinning this composite meaning are two beliefs. One (informed by religious and scientific thought) is in the existence of some kind of 'embryonic' or latent potential that is primed to develop/unfold into a 'full future form'. The second is in the capacity of humans 'to act intentionally to change existing conditions', including intervening to make something 'unfold' if it does not seem to be doing so adequately on its own.[22]

As discussed further in the next section, the tension between these two beliefs—what Cowen and Shenton call 'immanent development' and 'intentional development'—continues to stimulate debates about development today.[23] Core to this tension is whether intervention is necessary and, if so, how it relates to 'background' development or 'progress'. During the height of the colonial period, when development crystallised as an overarching policy framework, it was used not only as a tool for dispossession, extraction and settlement of new lands but also as the goal to justify these invasions. As Koponen explains, the general rationale was that: 'If indigenous people had left the resources of their countries undeveloped, their development was not only a right but also a duty of the colonialists'.[24] Animated by what Tania Murray Li refers to as 'the will to improve', the colonial project was justified by the assertion that it was of mutual benefit for the colonised as well as the colonisers.[25]

Yet, accumulating evidence of the lack of benefits enjoyed by the colonised, and by the working class 'back home', quickly strained the idea that colonisers were simply 'coaxing out' a natural potential in the world and that elite, capitalist development of natural resources was enough to generate benefits for local communities. Rather it became clear that extractive and industrialising processes were imposing an extreme cost on many local populations at home and abroad, including dispossessing them of their lands, undermining their livelihoods and eroding their health and survival.

As colonial governments struggled to develop some of their seized territories into proper countries, unemployment and inequities drove civil unrest in France and Britain, and critics such as Karl Marx and Frederick Engels deplored the inhumanities of industrialisation. In the midst of these struggles, social or 'human' development gradually emerged as something of a counterpoint or complement to 'economic development', although the latter remained the overall goal. In this way, the *practice* of development (the third leg of its composite meaning, mentioned above) was adjusted to better deliver on the *ideal* of development as being a kind of 'peaceful evolutionary change guided by conscious human action'.[26]

There are three important things to note about this history. The first is the link to universities. Education is deeply entwined with the notion of development and is similarly characterised by the tension between a belief in people's inherent latent potential (e.g. in a child) and the need for expert guidance and intervention (formal education, training) to ensure that potential is fully realised and directed towards what educators recognise as desirable ends. Unsurprisingly, formal education has long been a core human development intervention, motivated by a desire to both morally improve individuals and fulfil the labour needs of the economy. Clemente Abrokwaa argues that in colonial Sub-Saharan Africa, for example,

> Western education became the most sought after, important agent of social change within the different colonies. ... Western education became the index of development as well as the tool for measuring national and human growth.[27]

Also relevant to universities is the fact that research has had a central role in colonisation and associated conceptions of development, human civilisation and progress. As with education, science has functioned in colonialism as both means and end, tool and proof.[28] In the colonies, research institutes and associated networks and conferences became a major feature of imperial practices and circuits, helping fuel not only practical outcomes in local contexts such as large-scale irrigation but research in European-based universities.[29]

Beyond science's practical and symbolic role, social science also emerged as a key component of development. Indeed, according to some commentators, social science emerged as a field largely because it could purportedly understand and help shape society—that is, foster social development—as reliably as science could nature, making social science the complement to science and economic development.[30] In sum, the point is that modern universities and their contemporary challenges are, at least in settler colonial nations, partially a product of the ideal of development, which they are now being called upon afresh to support through the SDG agenda.

The second important point to note is that it is out of this development-centred context that the institution of the United Nations emerged. Contrary to what some people assume, the United Nations did not invent the idea of development—it is instead a product of it. That said, as we discuss below, the United Nations emerged in the post-World War period hand in hand with a new variant of development—what we now know as 'international development'. The implication is that the UN and development, including the SDGs, are closely linked, though not in a simple, linear fashion.

Third, the role of unemployment and the rumblings of civil unrest and 'violent revolution' in driving and challenging development in the past[31] begs the question of how development will feature in responses to the contemporary challenge of COVID-19 and its economic and social consequences. Although it has already been pointed out that the current crisis threatens to slow progress on achieving the SDGs, the historical pattern suggests it may also invigorate a rebooting and reworking of development, with implications for the SDG agenda.

International Development and Its Discontents

As indicated above, one of the thematic threads running through the SDGs is the notion of international development and its particular expression through the MDGs. Now a large and well-developed industry, international development emerged as a variant and continuation of colonial development in the post-World War period when the Bretton Woods agreement helped spark a new global imaginary—a new awareness of nations' integrated fates and fortunes. In a landmark speech in 1949, US President Truman called for a 'fair global development programme', not for charity's sake but because it would be mutually beneficial for all nations involved. As Truman put it, the poverty of under-developed nations 'is a handicap and a threat both to them and to more prosperous areas' (such as the US).[32]

The subsequent establishment of the International Bank of Reconstruction and Development and the United Nations (picking up where the inter-war League of Nations left off) helped to solidify this united view and put into practice international development flows of financial support from rich to poor through mutually obligated aid arrangements. In this way, development was rhetorically distanced from colonialism and 'reborn' as a modern global objective for all. Promoting the 'economic and social progress and development' of all people was written into the United Nations charter.[33]

Despite broad agreement on the need for international development, its implementation has been characterised by fierce debates over its actual direction or goal. As Koponen notes:

> Even if we speak, as the discourse of international development does, 'only' of economic and social development, its meanings cover a huge range: from modernisation to poverty reduction, from economic growth through increased productivity and the production of more-or-less-necessary gadgets to the fundamental values of a good life and the enlargement of human freedom.[34]

There has also been a long history of contestation over how development should be pursued and the degree and source of intervention, relative to leaving local contexts to 'develop' in a more immanent, bottom-up way. Adding to contestation and diversity in approaches is a pluralism of the groups involved. Some high-income countries such as Australia and the United States have backed away from the idea that there is a shared moral imperative to assist low-income countries, leading to an overall decline in the financial and political influence of nation states in international development. Other countries, namely China, have moved from being recipients to significant deliverers of foreign assistance.

Besides nation states, a diversity of increasingly professionalised and politicised actors now characterise international development. Philanthropic organisations/businesses such as the Gates Foundation, development professionals and companies (including those devoted to *assessing* development interventions in keeping with good governance standards), non-governmental organisations of all sorts, and large consortia such as CGIAR (formerly the Consultative Group for International Agricultural Research) now vie for influence in international development settings. This includes universities.

Further complicating the situation are three overlapping paradigms that have emerged over the last five decades as alternatives to conventional international development. These general alternatives and their arguments point to some important lessons from the past and a range of intellectual resources we can draw on to shape the future. As we argue in relation to the topics of sustainable development and resilient development below, appreciating the history of, and contestation around, international development helps us understand some of the criticisms levelled at the SDG agenda and thus tackle it in a more sophisticated and effective manner.

Neopopulist International Development

The first line of critique levelled at international development, advocates for an alternative 'participatory' or 'neopopulist' development approach. This approach maintains international development's commitment to deliberately transferring wealth from rich to poor, but argues against the classic top-down way in which associated development efforts are conducted, given the negative ways a significant proportion of international development efforts have affected local populations.[35]

Neopopulist international development advocates for development efforts to be largely led by local people, local knowledge and local human development priorities.[36] For example, rather than Western technologies being 'rolled out' in local agricultural contexts to try to increase others' food security or profits (as many colonial initiatives largely tried to do), the focus is on context-appropriate interventions and technologies—and indeed context-appropriate research and innovation, as we discuss in Chaps. 4 and 5. In terms of the SDG agenda, the neopopulist perspective on development is evident in the agenda's emphasis on localisation and participation, reflecting the long consultation process involved in formulating the SDG agenda. Nevertheless, numerous neopopulist critics voice legitimate reservations about the SDGs. As Belda-Miquel et al. (2019) note:

A key question is whether they can address structural problems in development aid policies and practices, such as the lack of accountability and coherence, unequal power relations, or depoliticisation.[37]

The authors conclude that: 'It seems that this will depend on how the agenda is adopted in the various territories as well as on the different interests at play'.[38] Their analysis of how the SDGs are being localised and implemented in the port city of Valencia on the Spanish coastline illustrates this point, highlighting the competing interpretations of the SDGs at work in the local context and the conflicting discourses involved in implementing them. As we return to below, this emphasis on the fact that the SDG agenda is not monolithic but is co-produced by actors as they interpret, debate and implement it within particular dynamic contexts is vital to appreciating the malleable nature of the SDG agenda, and key to why we believe the SDGs have positive and subversive potential.

It is useful here to consider the neopopulist criticism of the SDG agenda as itself a product of clashing worldviews. Being associated with the UN, the SDG agenda is interpreted by many people as a classic product of what Mary Douglas and colleagues would call a Hierarchical worldview.[39] This is a typically Western stance on the world that assumes and values the existence of a strong (hierarchical) social order, combined with a strong moral commitment to others. In their Cultural Theory worldview framework, which is based on empirical analysis of groups around the world, Douglas and colleagues refer to this as a 'strong grid' and 'strong group' orientation (see Fig. 2.1).

One of the three alternative worldviews in the resultant matrix shares the strong moral commitment to others but eschews the orientation to a strong grid. In contrast to the Hierarchical worldview's belief in the importance of formal leaders, professionals and experts, this Egalitarian worldview emphasises the role of the public, local communities and Traditional Owners. It also resonates strongly with the relational understanding of the world that characterises many Indigenous worldviews.

To some degree, the neopopulist critiques of international development and reservations about the UN-led SDG agenda expresses an Egalitarian worldview, and a related interpretation (arguably misinterpretation) of mainstream approaches (notably the UN) as too Hierarchical

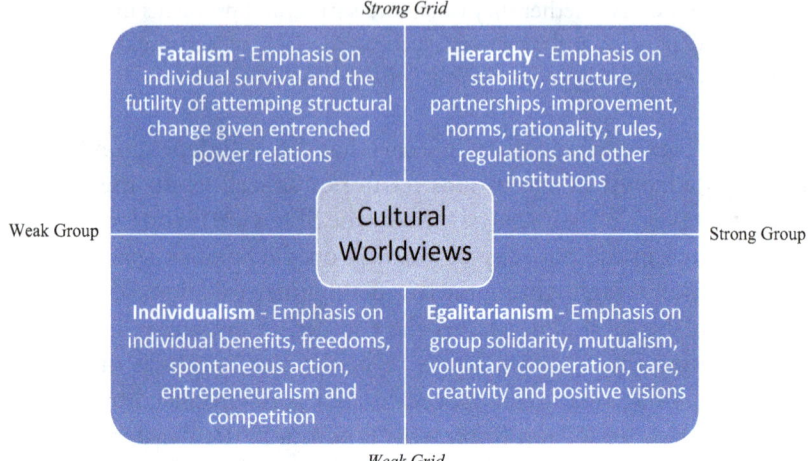

Fig. 2.1 The four worldviews of Cultural Theory. (Adapted from the work of Mary Douglas and colleagues. Figure from https://www.dustinstoltz.com/blog/2014/06/04/diagram-of-theory-douglas-and-wildavskys-gridgroup-typology-of-worldviews)

in nature. It is a concern that we return to below, along with a further discussion of worldviews and how they help us appreciate characteristics of the SDG agenda.

Neoliberal Development

International development has been more strongly critiqued from a second alternative approach—neoliberal development—which dismisses the whole notion of international development and even rejects intentional development by states at all. As the name suggests, neoliberal development is a product of the pro-Capitalist market, neoliberal governance approach that rose to international prominence in the 1970s.[40] At the time, the end of the Cold War meant that international development was losing its status as a tool of soft power within larger geo-political struggles. The combined rise of neoliberal economic policies and the reduced importance of international development for national geo-political agendas meant that the whole premise of intentional development—and especially intentional development via state aid or 'welfare' programmes—was called into question. Although the first two arms of the composite meaning of development described above remained relatively intact (i.e. the belief in development as a general goal and process), the third meaning (intentional human intervention to engender the process and achieve the goal) was largely rejected.

The result was acknowledgement that many countries of the world remain 'under-developed' (e.g. with levels of child mortality or hunger far above the global average), but strong resistance to addressing this using government welfare and the so-called developmental state.[41] Instead the World Bank, International Monetary Fund, US Treasury and others established a set of free market policies (the Washington Consensus) to guide development. These include Structural Adjustment Programs that replaced development aid with the 'carrot' of financial loans to push recipient national governments to liberalise their governance structures and cultures.

Ongoing today, the goals of these programmes include bringing countries more fully into the global economy and reducing barriers to foreign trade, reforming their governance structures and processes to better meet modern standards and reduce corruption, and implementing specific development projects such as large infrastructure projects or microfinance to stimulate entrepreneurial behaviour. Over time the neoliberal perspective has widely popularised the idea that simply giving humanitarian aid to people in need distorts markets and disincentivises individuals and nations from helping themselves.[42]

Combined with the Global Financial Crisis, an upshot of the neoliberal turn is that since the 1960s total Official Development Assistance payments from wealthy nations for international development have fallen as a proportion of donor country's income, with only Sweden, Denmark and Norway consistently meeting the agreed UN target of 0.7% of Global Net Income.[43] This drop in financial assistance is despite a partial retreat from the harsh neoliberal policies of the 1980s triggered by unavoidable evidence of the regressive social and environmental consequences of one-dimensional Structural Adjustment Programs. Some commentators even declared neoliberalism 'dead' after the 2000–2015

Millennium Development Goals helped 'revive' intentional, international development.[44] As others have pointed out, however, if neoliberalism has faded at all, it is far from dead and remains in a zombie state.[45]

Although the SDG agenda initially helped further bolster foreign aid by encouraging many wealthy nations to restate their commitment to intentional international development, and the overall amount donated in 2016 reached a record high, more recently this commitment has started to waver, thanks in part to the rise of more nationalistic and neoliberal policies in countries once prominent in international development assistance such as the UK, US and Australia. Emma Mawdsley and colleagues characterise the current regime as 'retroliberalism'—one in which there is a stated commitment to 'shared prosperity' but also 'a return to explicit self-interest designed to bolster private sector trade and investment' and 'aid programmes … increasingly functioning as "exported stimulus" packages'.[46]

Meanwhile, the amount of funding needed for international development continues to rise as humanitarian crises increase in number and length, due in part to the cascading effects of climate change. In 2019 it was reported that half-way through the year 'humanitarian organisations had still received less than a third of money—27%—they needed to provide relief to people affected by crises worldwide'.[47] Since then, COVID-19 and its flow-on effects have compounded the problem. The situation is so serious that New Zealand development scholars John Overton and Warwick Murray assert that 'despite a global commitment to the Sustainable Development Goals, we are at a point where the very notion of aid is being questioned and its future is uncertain'.[48]

Over the last four decades the roll back of social welfare programmes run by recipient countries and international development assistance from wealthy countries has created a vacuum that civil society and a plethora of development non-governmental organisations have had to step into.[49] To some extent this has generated a window of opportunity for more participatory, local-based initiatives to thrive in keeping with the neo-populist critique of top-down international development mentioned above. It has also precipitated a turn to the private sector and philanthropies to try to fill the financial gap. This has deepened the influence of capitalism on international development by not only involving a new range of global corporations in international development work but by stimulating the financialisation of the development sector, adding in a level of complicated financial instruments that businesses far removed from development can participate in and profit from.[50]

From a neoliberal perspective, the UN is often critiqued as hopelessly old-fashioned, bureaucratic and naïve. That is, it is once again criticised as too Hierarchical—this time from the Individualistic worldview (not Egalitarian worldview) that characterises capitalism and the neoliberal ideal. Free of a moral commitment to help, Individualists critically appraise international development in terms of opportunity and self-benefit. In this light, international development is potentially another arm of business, offering new markets, financial assets, labour and opportunities to demonstrate Corporate Social Responsibility in order to legitimate other business activities.

The role of the SDG agenda is ambiguous here. On the one hand, its overlap with the MDGs means it is often interpreted as a nation state and altruistic issue, a continuation of UN moralising of a sort that many have tuned out to or regard dismissively, reflecting to some degree the normalisation of an Individualist outlook on the world. On the other hand, the business community is far more explicitly involved in the SDGs than the MDGs. The private sector had a powerful influence on the design of the agenda and businesses are specifically charged with helping to implement it—both in terms of helping cover the trillions (1.5–2.5% of global GDP) estimated to be needed per year to cover implementation costs[51] and in terms of driving specific initiatives.[52]

Post-Development

We come then to the third and strongest line of criticism directed at international development. The 'post-development' paradigm calls into question the entire modernist premise of development—not just international development but earlier colonial development as well as the sustainable development approach discussed below. As Wolfgang Sachs famously wrote in the introduction to *The Development Dictionary* in

1992: 'The last 40 years can be called the age of development. This epoch is coming to an end. The time is ripe to write its obituary.'[53]

Later in the Dictionary, Gustavo Esteva similarly describes 'development' as an 'unburied corpse',[54] while more recently, Eduardo Gudynas argues that development is 'a zombie concept, dead and alive at the same time'.[55] All the talk of death and zombies indicates that, unlike the neo-populist international development approach described above, post-development does not 'intend to improve the attempts to bring about "development" but questions "this very objective"'.[56] As Aram Ziai outlines, from a post-development perspective, development is:

1. an ideology of the West, promising material affluence to decolonising countries in Africa and Asia in order to prevent them from joining the communist camp and maintaining a colonial division of labour ....
2. a failed project of universalising the way of life of the 'developed' countries on a global scale which has for the overwhelming majority of affected people led to the 'progressive modernization of poverty' ....
3. a Eurocentric and hierarchic construct defining non-Western, non-modern, non-industrialised ways of life as inferior and in need of 'development' ....
4. an economic rationality centred around accumulation, a capitalist logic of privileging activities earning money through the market (and disvaluing all other forms of social existence), and the idea of the *Homo economicus* (whose needs for consumption are infinite).
5. a concept that legitimises interventions into the lives of people defined as 'less developed' as justified in the name of a higher, evolutionary goal or simply the common good defined by people claiming expert knowledge.[57]

So how does the SDG agenda look from this perspective? Ziai raises this question explicitly, asking whether the new agenda has 'provided a rejuvenating cure' or whether it is 'only the last in a long line of cosmetic surgeries designed to let its object appear fresh and vigorous, but unable to mask the signs of decay?'[58] As his tone suggests, Ziai is unimpressed by the SDG agenda, as are critical scholars such as Heloise Weber who point to specific limitations such as its promotion of capitalism and free trade.[59]

2 Sustainable Development in the Anthropocene 55

From this perspective the SDG agenda is read as not just Hierarchical but also Individualist, a kind of Jekyll and Hyde monster that reflects the fact that neoliberalism is equally the progeny of government as business. As seen in Ziai's list above, the SDGs are interpreted as yet another elitist, bureaucratic, imperial endeavour, if not a calculated and dangerous bid at neo-colonialism.

Significantly, however, there are at least two variants of the post-development paradigm which understand the problem from different worldviews and thus differ in their preferred response. First, there are those characterised by deep cynicism about not only development's specific ambition of improving the human condition but grand ambitions of any sort. At work here, we suggest, is a Fatalist worldview (Fig. 2.1) which understands the world as deeply unjust and everyone as only out for themselves. From this perspective the SDG agenda is a ludicrous initiative and/or a poorly described grab by entrenched interests for yet more power. Given their deep despair and apathy about the world, this cynical camp does not offer any suggestions as to what could replace development. Rather, as discussed below in terms of the Anthropocene, the focus is just on coming to terms with the end of the human story.

The second variant of post-developmentalism is more action-oriented. Here there is no question that the whole paradigm of development needs overhauling, but there *is* a belief in the capacity for such transformative change. In particular, there is a burgeoning of scholarship and practice around identifying, celebrating and experimenting with specific, tangible alternative models. In keeping with an Egalitarian worldview, these alternatives often highlight the value of marginalised philosophies and perspectives, such as the *Buen vivir* ('living well') framework of Indigenous groups in Ecuador or the degrowth paradigm in economics.[60]

At the same time, there is some reflexivity in this variant of post-developmentalism about the irony that some of the most strident advocates *for* conventional development—for example, development projects that improve sanitation, incomes, health care, good governance—are from those living in the 'developing world' contexts that post-development advocates claim to be representing or at least protecting from development.[61] Critics of post-development call out post-development scholars for declaring development 'over' when they largely do so from positions of privilege that have been enabled by that very development, yet deny such benefits, meaning that they are effectively 'pulling up the ladder' after them.

Amplifying this scrambling of positions is an emerging shift towards less binary 'for/against' thinking. Even Wolfgang Sachs, the so-called father of post-development, recognises that the SDG agenda is an assemblage of many worldviews, ideologies and agendas and cannot be easily boxed as bad or good.[62] Although he points out that the SDG agenda is less progressive than Pope Francis's remarkable 2015 Encyclical letter *Laudato Si* (which resonates strongly with *Buen vivir*), he does see real potential in the SDGs—as do we. Overall, the point is that post-development, as with international development, is characterised by a tense combination of, on the one hand, mounting evidence of the vital importance of its underpinning concern with development and equality and, on the other hand, keen awareness of and growing frustration with the limitations of dominant development approaches.

Sustainability and the Anthropocene

If development is an unfolding of human potential and ongoing improvement of human society, *sustainable* development is an effort to guide it in such a way that it fosters, not erodes, our long-term environmental enabling conditions and so can be sustained over time. As a concept, sustainable development was institutionalised and popularised with the 1987 report *Our Common Future* (the Brundtland report) by the World Commission on Environment and Development, an international working group set up by the UN General Assembly in 1983 to propose strategies 'for achieving sustainable development to the year 2000 and beyond'.[63]

Our Common Future defined sustainable development as that which 'meets the needs of the present without compromising the ability of future generations to meet their own needs'. In doing so, it helped crystallise a new global sensibility, future-orientation and moral ideal. More specifically, it addressed a number of emerging concerns about development, beginning with the need for a more integrated approach. Discussing *Our Common Future*, John Dryzek asserts that:

> Its main accomplishment was to combine systematically a number of issues that have often been treated in isolation, or at least as competitors: development, global environmental issues, population, peace and security, and social justice both within and across generations.[64]

The concept of sustainable development has also helped illuminate numerous other realities: the need to understand development as a continuous process involving all countries and all parts of the Earth, not just colonies or places of international development intervention; the need to reshape development to better fit the limits of the planet; and the need to attend more carefully to reproductive as well as productive processes, including those that care for, maintain and repair the world.

Approaching development in more global terms and establishing a 'new international order' had already been flagged thanks to the debates about international development discussed above. For example, the Independent Commission on International Development released an influential report, *North-South: A Programme for Survival* (the Brandt report), in 1980 that argued strongly for the rights of those in the Global South to greater redistribution of wealth from the Global North (given the dependency relations the latter had established) and to a greater say in 'international political and economic affairs'. *Our Common Future* built on and diverged from this language of rights and responsibilities by taking it as a given that all countries were equal and focusing instead on the question of mutual interests.[65] Relative to other approaches, sustainable development emphasises the need for coordinated action by actors across the world at all levels, 'motivated by the public good'.[66]

In the approximately three decades between *Our Common Future* and the SDG *Transforming Our World* agenda, a lot has happened, but the outcomes envisaged by the authors of the Brandt and Brundtland have not been realised. Deep socioeconomic and political inequalities persist, and while many alternative approaches to sustainable development have been tried and hotly debated, the planet itself has also heated up and many other environmental indicators have continued to decline. It has become clearer than ever that we are transforming the Earth itself, not just because the list of individual environmental problems is lengthening but because their complex interactions are altering how the Earth System itself functions, pushing us into what is now known as the Anthropocene.[67]

It has also become clearer that the interpretations of 'sustainable development' that have come to prominence since *Our Common Future* have failed to grasp or address the challenge. Some definitions of sustainable development are a lot more radical than others. Systems thinkers Donella Meadows and colleagues, for example, endorse *Our Common Future* for what they see as its implicit questioning of the paradigm of economic growth, in keeping with their own global systems analyses (e.g. the *Limits to Growth* report) which point to highly disruptive physical feedbacks (e.g. climate change, resource depletion and degradation) increasingly undermining economies and societies unless consumption and production processes are contained.[68]

In contrast, the dominant ways in which sustainable development has been defined and enacted (at least until recently) have presumed that economic growth is not only compatible with sustainable development but a requirement of it. These mainstream approaches to sustainable development are generally based on 'weak sustainability'—the idea, originally advanced by economist Robert Solow,[69] that economic development is sustainable, and nature can be squeezed hard as long as capital is reinvested in productive capabilities such as technological replacements for natural resources or processes.

Today, there is a dawning realisation that what is needed is not only a sufficient supply of resources, nor even the preservation of irreplaceable, non-commensurable natural resources (known as 'strong sustainability'), or even of patches of nature for its own sake.[70] Instead, thanks to advances in ecological, resilience and Earth System science, it is increasingly apparent that to protect 'our common future' we need to maintain the *functional integrity* of the Earth System itself.[71] This is an exceedingly more complex endeavour—one that extends far beyond the purview of the 'environment sector' to implicate all sectors, all organisations, all disciplines.

It is also one deeply complicated by the emergence of escalating feedbacks of the sort that *Limits to Growth* warned of fifty years ago. In systems terms, climate change, deforestation and related Anthropocene issues are starting to erode the planet's negative feedback loops (i.e. self-correcting mechanisms such as increased uptake of carbon dioxide by vegetation in conditions of high atmospheric carbon dioxide) and generate new positive feedback loops (self-amplifying mechanisms, such as wildfire begetting more wildfire as vegetation evolves to become more flammable and smoke produces greenhouse gases and worsens climate change).[72]

As a result, the planet is becoming less stable and predictable in its function. Combined with more localised pressures such as urbanisation, as well as the long supply chains, transnational circulations and interdependencies of knotted global systems, global risks are escalating in number and magnitude. A 2014 comparison of contemporary data with the dozen scenarios the *Limit to Growth* report modelled suggest that the world is tracking what was aptly named the Business-as-Usual scenario. Concerningly, it projects feedbacks and resource scarcities that increasingly disrupt economies and severely impact human populations.[73]

Planetary sustainability and resilience, like the concept of development, encompass all nations, sectors, individuals and actions. Which is one reason that universities are inescapably part of it and are crucial to addressing it. To understand the reciprocal role of universities within the contemporary sustainable development challenges presented by the Anthropocene and its uneven expression in the SDG agenda, we outline three key aspects in subsequent chapters: the need to face unsustainability; the need for resilience, adaptation and experimentation; and the need for maintenance, repair and regeneration. Each helps address the inevitable question of 'what should we do?'. In addressing this question, we aim to provide further insight into our motivations for writing this book and why we believe that the SDGs are a flawed but valuable tool for progressing the positive transformational change needed, including through universities as the next chapter elaborates on.

Notes

1. UNDP (undated) Sustainable Development Goals. https://www.undp.org/content/dam/undp/library/corporate/brochure/SDGs_Booklet_Web_En.pdf
2. UN (2015) Millennium Development Goals Report 2015. https://www.un.org/millenniumgoals/2015_MDG_Report/pdf/MDG%202015%20rev%20(July%201).pdf
3. Freistein, K., Mahlert, B. (2016) The potential for tackling inequality in the Sustainable Development Goals. Third world Quarterly 37, 2139–2155.
4. Sachs et al. (2020) Sustainable Development Report, SDSN https://s3.amazonaws.com/sustainabledevelopment.report/2020/2020_sustainable_development_report.pdf
5. UN (2020) *Emission Gap Report 2020* accessed on https://www.unep.org/emissions-gap-report-2020
6. Op cit.
7. Pyne, S. (2020) From Pleistocene to Pyrocene: fire replaces ice. *Earth's Future* 8, e2020EF001722.
8. NOAA National Centers for Environmental Information, *State of the Climate: Global Climate Report for Annual 2019*, published online January 2020, accessed February 12, 2021 from https://www.ncdc.noaa.gov/sotc/global/201913
9. NASA (2020) *Global Climate Change—Vital Signs of the Planet. Global Temperature.* Latest Average Annual Anomaly. https://climate.nasa.gov/vital-signs/global-temperature/ Accessed February 10, 2021.
10. UNEP (2019) *Emissions Gap Report* 2019.
11. IPCC (2018) Summary for Policymakers. In: *Global warming of 1.5°C. An IPCC Special Report on the impacts of global warming of 1.5°C above pre-industrial levels and related global greenhouse gas emission pathways, in the context of strengthening the global response to the threat of climate change, sustainable development, and efforts to eradicate poverty* [V. Masson-Delmotte, P. Zhai, H. O. Pörtner, D. Roberts, J. Skea, P. R. Shukla, A. Pirani, W. Moufouma-Okia, C. Péan, R. Pidcock, S. Connors, J. B. R. Matthews, Y. Chen, X. Zhou, M. I. Gomis, E. Lonnoy, T. Maycock, M. Tignor, T. Waterfield (eds.)]. World Meteorological Organization, Geneva, Switzerland, 32 pp.

2 Sustainable Development in the Anthropocene 61

12. Field, C.B., Barros, V.R., Mastrandrea, M.D., Mach, K.J., Abdrabo, M.-K., Adger, N., Anokhin, Y.A., Anisimov, O.A., Arent, D.J., Barnett, J., (2014) Summary for policymakers, *Climate change 2014: impacts, adaptation, and vulnerability.* Part A: global and sectoral aspects. Contribution of Working Group II to the Fifth Assessment Report of the Intergovernmental Panel on Climate Change. Cambridge University Press, pp. 1–32.
13. IPBES Global Assessment Report (2019) Summary for Policy Makers, p. 10–15.
14. Ibid., p. 10.
15. Ibid., p. 10, 15.
16. Sachs et al. (2020) p. 25.
17. Gabay, C., Ilcan, S. (2017) Leaving No-one Behind? The Politics of Destination in the 2030 Sustainable Development Goals. *Globalizations* 14, 337–342.
18. Fankhauser, S. and Schmidt-Traub, G. (2011) From adaptation to climate-resilient development: the costs of climate-proofing the Millennium Development Goals in Africa. *Climate and Development* 3, 94–113.
19. Koponen, J. (2020) Development: History and Power of the Concept. *Forum for Development Studies* 47, 1–21. P. 5.
20. Massey, D. (1999) Spaces of politics. In: D. Massey, J. Allen and P. Sarre (eds) *Human Geography Today.* Polity Press, Cambridge. Pp. 279–94.
21. Koponen, J. (2020) Development: History and Power of the Concept. *Forum for Development Studies* 47, 1–21. P. 5.
22. Ibid., p. 7.
23. Cowen, M.P., Shenton, R. (2003) *Doctrines of Development.* Routledge, London. Hart, G. (2010) D/developments after the Meltdown. *Antipode* 41, 117–141.
24. Ibid., p. 9.
25. Murray Li, T. (2007) *The will to Improve: Governmentality, development, and the practice of politics.* Duke University Press, Durham.
26. Koponen, J. (2020) Development: History and Power of the Concept. *Forum for Development Studies* 47, 1–21. p. 14.
27. Abrokwaa, C., (2017) *Colonialism and the development of higher education, Re-thinking Postcolonial Education in Sub-Saharan Africa in the 21st Century.* Brill Sense, p. 201–220. P. 206.
28. See for example Gascoigne, J., Tranter, N. (1998) *Science in the service of empire: Joseph Banks, the British state and the uses of science in the age of*

revolution. Cambridge University Press. Dubow, S. (2006) *A Commonwealth of Knowledge: Science, Sensibility, and White South Africa 1820–2000.* Oxford University Press on Demand.
29. Harrison, M. (2005) *Science and the British empire.* Isis 96, 56–63.
30. See for example Wallerstein, I. (1984) The Development of the Concept of Development. *Sociological Theory* 2, 102–116. Gilmartin, D. (1994) Scientific Empire and Imperial Science: Colonialism and Irrigation Technology in the Indus Basin. *The Journal of Asian Studies* 53, 1127–1149.
31. Op cit.
32. Koponen, p. 16.
33. Op cit.
34. Ibid., p. 3.
35. See for example Murray Li, T. (2007) *The Will to Improve: Governmentality, development, and the practice of politics.* Duke University Press, Durham. Barber, M., Bowie, C. (2008) How international NGOs could do less harm and more good. *Development in Practice* 18, 748–754. Neef, A., Singer, J. (2015) Development-induced displacement in Asia: conflicts, risks, and resilience. *Development in Practice* 25, 601–611.
36. Cowen and Shenton 2003.
37. Belda-Miquel, S., Boni, A., Calabuig, C. (2019) SDG Localisation and Decentralised Development Aid: Exploring Opposing Discourses and Practices in Valencia's Aid Sector. *Journal of Human Development and Capabilities* 20, 386–402. P. 386.
38. Ibid.
39. M. Douglas (1999) Four cultures: The evolution of a parsimonious model. *Geojournal* 47, 411–415. Douglas, M., Wildavsky, A. (1983) *Risk and culture: An essay on the selection of technological and environmental dangers.* Univ of California Press. Thompson, M., Ellis, R., Wildavsky, A. (1990) Cultural theory. Westview Press.
40. See for example Harvey, D (2006) *A Brief History of Neoliberalism*, Oxford, Oxford University Press.
41. Fine, B. (1999) The developmental state is dead—long live social capital? *Development and Change* 30, 1–19.
42. Overton, J., Murray, W.E. (2020) *Aid and Development.* Routledge. Weyland K., 1996: Neopopulism and neoliberalism in Latin America: Unexpected affinities. *Studies in Comparative International Development,* 31, 3–31.

43. Inter-Agency Task Force on Financing for Development (2016) *Official Development Assistance. Issues Brief.* https://www.un.org/esa/ffd/wp-content/uploads/2016/01/ODA_OECD-FfDO_IATF-Issue-Brief.pdf
44. Murray, W.E., Overton, J.D. (2011) Neoliberalism is dead, long live neoliberalism? Neostructuralism and the international aid regime of the 2000s. *Progress in development studies* 11, 307–319.
45. Murray, W.E., Overton, J.D. (2011) Neoliberalism is dead, long live neoliberalism? Neostructuralism and the international aid regime of the 2000s. *Progress in development studies* 11, 307–319.
46. Mawdsley, E., Murray, W.E., Overton, J., Scheyvens, R., Banks, G. (2018) Exporting stimulus and "shared prosperity": Reinventing foreign aid for a retroliberal era. *Development Policy Review* 36, O25–O43.
47. https://www.theguardian.com/global-development/2019/jul/16/alarming-shortfall-foreign-aid-worlds-biggest-crises
48. Overton, J and Murray, W (2020) *Aid and Development*, London, Routledge.
49. Fine, B (1999) The development state is dead—long live social capital? *Development and Change,* 30(1), p. 1–19.
50. Järvelä, J., Solitander, N., (2019) The financialization and responsibilization of development aid, in: Lund-Thomsen, P., Wendelboe Hanson, M., Lindgreen, A. (Eds.), *Business and Development Studies: Issues and Perspectives.* London, New York, p. 100–122.
51. https://irp-cdn.multiscreensite.com/be6d1d56/files/uploaded/151112-SDG-Financing-Needs-Summary-for-Policymakers.pdf
52. Scheyvens, R., Banks, G., Hughes, E. (2016) The Private Sector and the SDGs: The Need to Move Beyond 'Business as Usual'. *Sustainable Development* 24, p. 371–382.
53. Sachs, W., ed. (1992) *The Development Dictionary. A Guide to Knowledge as Power*, London: Zed Books.
54. Esteva, G. (1992) "Development." In *The Development Dictionary. A Guide to Knowledge as Power*, edited by Wolfgang Sachs, London: Zed Books. Pp. 6–25.
55. Gudynas, E. (2011) Buen Vivir: today's tomorrow. *Development* 54, 441–447. P. 442.
56. Ziai, A. (2017) Post-development 25 years after The Development Dictionary. *Third world quarterly* 38, 2547–2558.
57. Ibid. 2017, p. 2547–48.
58. Ibid., p. 2547.

59. Weber, H. (2017). "Politics of 'Leaving No One Behind': Contesting the 2030 Sustainable Development Goals Agenda." *Globalizations* 14(3), p. 399–414.
60. Bendix, D. (2017). "Reflecting the Post-Development gaze: the degrowth debate in Germany." *Third world quarterly* 38(12), p. 2617–2633. Acosta, A. (2017). "Living Well: ideas for reinventing the future." *Third world quarterly* 38(12), p. 2600–2616.
61. For example see Matthews, S. (2017). "Colonised minds? Post-development theory and the desirability of development in Africa." 38(12), p. 2650–2663.
62. Sachs, W. (2017). "The Sustainable Development Goals and Laudato si': varieties of Post-Development?" *Third world quarterly,* 38(12), p. 2573–2587.
63. Brundtland, G. (1987). *Report of the World Commission on Environment and Development: Our Common Future.* United Nations General Assembly document A/42/427.
64. Dryzek, J.S. (2012) The Politics of the Earth: Environmental discourses, 3rd ed. Oxford University Press. P. 150.
65. Thompson, P. B. (2010). *The Agrarian Vision: Sustainability and Environmental Ethics.* Lexington, The University Press of Kentucky.
66. Dryzek (2012), p. 160.
67. Steffen, W., Rockström, J., Richardson, K., Lenton, T.M., Folke, C., Liverman, D., Summerhayes, C.P., Barnosky, A.D., Cornell, S.E., Crucifix, M., Donges, J.F., Fetzer, I., Lade, S.J., Scheffer, M., Winkelmann, R., Schellnhuber, H.J. (2018) *Trajectories of the Earth System in the Anthropocene.* Proceedings of the National Academy of Sciences.
68. Dryzek (2012). See also Meadows, D.H., Meadows, D.L., Randers, J., Behrens, W.W. (1972) *The Limits to Growth; a Report for the Club of Rome's Project on the Predicament of Mankind.* Universe, New York.
69. Solow, R.M. (1992) *An Almost Practical Step Toward Sustainability*; Resources for the Future: Washington, DC, USA.
70. Norton, B. G. (2005). *Sustainability: A philosophy of adaptive ecosystem management*, University of Chicago Press.
71. Ibid.
72. Meadows, D. (1999). *Leverage Points: places to intervene in a system.* http://www.sustainer.org/pubs/Leverage_Points.pdf The Sustainability

Institute. On the increasing flammability of the planet due to climate change, see Pyne (2020) above.

73. Turner, G. (2014) *Is global collapse imminent? An updated comparison of the Limits to Growth with historical data*. Melbourne Sustainable Society Institute Research paper no. 4, University of Melbourne. Pp. 21. https://sustainable.unimelb.edu.au/publications/research-papers/is-global-collapse-imminent. Accessed February 5, 2021.

3

The Role of the University in Society

The New Normal

We have been warned that global risks are escalating and there is no going back. This was the general consensus to questions posed by the Times Higher Education webinar panel: 'Has Covid-19 changed universities forever?', and 'What new iterations of the university might emerge from the rubble?'.[1] The 'new normal' is the latest moniker for the state of higher education: one that reflects universities grappling with the seismic shifts to core operations and structures of universities affected by the COVID-19 crisis. More broadly within this twenty-first-century context, the identity and purpose of the university—like society itself—is in a state of systemic flux, crisis and change. As Isaac Kamola writes, 'It is important to remember that all universities are always already multiple, with many histories, and many crises'.[2]

The isolation and pain that is being felt across universities is both a reality and metaphor for universities and the higher education sector writ large. COVID-19 is just one of many crises and calamities re-shaping the ideal of 'the University' and real universities on the ground. The closure of university campuses across the world to staff and students in lock-down has forced core activities such as learning and teaching, research and engagement, management and administration to increasingly take place in a hybrid mode or online. This is underpinned by a dispersed and largely invisible web of highly unequal private spaces, from the crowded kitchen tables of casual tutors to the palatial home offices of university executives. Meanwhile, the precariousness of the university funding model and over-reliance of many institutions on international student enrolments have re-opened important questions about the contemporary role of the university. The critical question is how the crises faced-by universities relate to longer standing critiques of them.

Among the many tensions and fault lines in contemporary universities intensified by the pandemic, three stand out.

First, the *emphasis on using universities to drive economic growth and development* as a route to recovery has amplified the contested role of universities as a tool in the state's economic toolkit and the shaping of universities as corporations providing research and educational services for a fee. Pandemic-induced job losses in some universities have illuminated the widening gap between executive salaries and large property portfolios on the one hand, and increasingly precarious and vulnerable staff and students on the other hand. Meanwhile, although some governments have responded with generous funding injections to higher education, in other contexts the challenges of increasing numbers of students and staff-student teaching ratios, and the concomitant reduction of public funding per student over the lifetime of the study course, is pushing universities further on their quest for new income sources and resources—and thus towards wealthy industry partners, for-profit operating models and an economic framing of higher education.

Second, pandemic experiences highlight *the linkages and blurred boundaries between universities, and between universities and the rest of society*. In some cases, this has seen universities pulling together and advocating as a coalition, whilst in others it has heightened competition between them. All universities are juggling the dynamic effects of the pandemic and its repercussions on their local context *and* on the far-flung sites and usually international flows of students, staff and resources that they often rely on. Together these geographic relations illuminate the physical forms and embeddedness of universities in multiple places. This has reinforced the inescapable connections between universities and domestic environments, with the unequal effects of home schooling and domestic care responsibilities on different staff and students, for instance, highlighting the pervasive nature of gender inequalities in the societies that universities are part of.[3]

Third, the pandemic has reinforced growing concern within universities about the *deepening global challenges facing humanity and the planet.* With mounting evidence that the world's current trajectory is unsustainable, it is increasingly clear that universities cannot externalise the costs and risks of existing ways of doing things or downplay such issues as irrelevant, insignificant or merely interesting topics for discussion. It is telling that more and more universities, disciplines, higher education networks and student groups are joining others in declaring a climate emergency, calling for urgent climate action, and even progressing such action within their own institutions and organisations. This includes eighty-seven universities that collectively committed to climate action as part of their involvement in the SDG Accord—a voluntary agreement that represents 'the university and college sector's collective response to the global goals'.[4]

A major challenge for climate action in the sector is the fact that most sustainability initiatives within universities (which climate change is still somewhat erroneously framed as) end up as side-lined in separate units, strategies and policies, unable to influence the core business decisions or culture of the institution, and are thus severely limited in ambition and effectiveness. As Claudia Zwar and Simon Lancaster (2020) write about the UK situation:

> Universities are natural leaders in combating climate change and the flurry of recent environmental targets is overdue. But there is a real risk that without placing sustainability in a broader way of thinking about success in higher education, these climate strategies create more hot air.[5]

Fortunately, there seems to be a growing desire to genuinely make all aspects of higher education 'climate compatible' by placing greenhouse gas emissions reduction and climate change adaptation at the heart of strategic decision-making in the sector. One of the results is that usually implicit models of success, such as conservative notions of research excellence and conventional economic growth (discussed above), are coming under scrutiny. In particular, climate change has exposed some of the costs, risks and myths of the sector's global mobility ideal, including the assumption that frequent travel fosters personal success.[6] Rising awareness of the highly polluting, risky and often unnecessary nature of long-distance international travel means that its disruption during the COVID-19 pandemic can be less easily dismissed as unforeseeable or framed as a problem we need to simply bounce back from. Instead, there is an appetite for rethinking some of the fundamentals of the global higher education sector and revisiting the purpose and role of universities.

Given this context, how will universities engage with the Sustainable Development Goals (SDGs)t as a transformational vision for achieving social equity and environmental sustainability over the coming decade? Universities have the opportunity and capacity to lead on SDG innovation across their four primary functions of research, education, external leadership, and operations and governance, and many have started trying to do so. But it is unclear if action to date is driving fundamental change or remains a side project or broad ambition. The SDGs demand many things from higher education beyond business-as-usual.

In Chap. 2, we outlined how the idea of development and its roots in modernity and colonialism are central to both the Sustainable Development Goals and the higher education sector. This chapter builds on this to position the SDG agenda as part of a shift in expectations about the role of the university. As illustrated schematically in Fig. 3.1, universities are increasingly understood not as ivory towers, oddities or innocents, but as deeply embedded and engaged with the world, including the crises, disruptions and unwanted changes that characterise them.

Moving beyond the nationalistic and individualistic competitor mindset, the SDGs encourage universities to heed the global call to action. As the world turns with the pace of new economies and technologies, and grapples with the challenges of intergenerational equity and justice, a global pandemic and planetary tipping points, the university is emblematic of humanity's quest for survival. The recent COVID-19 crisis is the symptom and not the root cause of the modern university in crisis: the

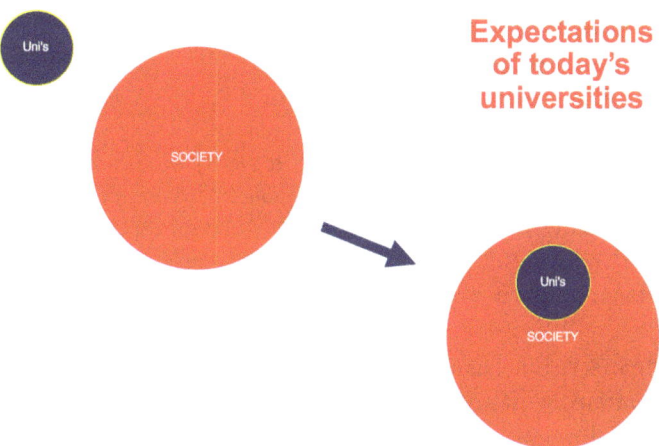

Fig. 3.1 Embedded and engaged—expectations of twenty-first-century universities (Source: Authors)

catalytic moment reinforcing the need for wider systemic and transformative change reflected in the SDGs.

The Idea of the University

Within the twenty-first-century context, 'is it possible to come forward with an idea of the university, that has a measure of feasibility, anchored in the real world and expressive of some hope and a measure of optimism?'.[7] This is how Ronald Barnett introduces his utopian vision of 'the Ecological University'. To rescue the university from impotence and defeatism, he argues, we must reconceive the university and its place in the world and the earth anew. We will return to the idea of the Ecological University later in the chapter, but first explore some of the ambitions and criticisms that have shaped the contemporary higher eductaion sector paradox whereby, 'the university is most needed at a moment when it is most in peril'.[8]

In 1852 'the idea of the university', according to John Henry Newman, was of a community of thinkers engaged in a broad liberal education. For Newman, writing over 150 years ago, the university was a place where:

> students could come from every quarter for every kind of knowledge, for the communication and circulation of thought, where inquiry is pushed forward, discoveries verified and perfected, and error exposed, by the collision of mind with mind, and knowledge with knowledge creating a pure and clear atmosphere of thought, which the student also breathes.[9]

Newman's vision has since been critiqued for being both elitist and anti-utilitarian in nature, in contrast to the vocational skills or professional accreditation that many universities now emphasise to serve specific (i.e. far more narrow) external purposes.[10] However Newman's ideas of a broad liberal education are being revisited during the COVID-19 crisis. Fostering a 'community of thinkers' in which students learn 'to think and to reason and to compare and to discriminate and to analyse'[11] in order to better address the uncertain future seems more important than ever. As Sophia Deboick wrote in *The Guardian* more than a decade ago, Newman's ideas speak to 'the soul' of the university, 'reminding us that the university has a greater role than just doling out qualifications—that of shaping the whole individual'. She continues: 'Newman's thought may usefully guide us as crucial decisions are made about the future of our universities'.[12]

The liberal notion of holistic education advocated by Newman built on earlier work by Prussian philosopher Wilhelm von Humboldt. His idea of the modern university influenced nineteenth-century European and later the elite American universities such as Harvard and Stanford. The Humboldtian Model was embedded in the enlightenment ideals of developing the autonomous individual *and* world citizen whereby 'knowledge is power, and education is liberty'. This included a focus on building learning and knowledge through the integration of arts and science with research; allowing students the freedom to choose their own studies; organisational independence within a system of state-based funding; and an emphasis on intellectual freedom. Critical thinking skills

3 The Role of the University in Society 73

were key to Humbolt's vision of the university: 'to inquire and to create—these are the grand centres around which all human pursuits revolve'.[13]

In the twentieth century this vision of universities designed to educate a small, exclusive group of scholars destined for the civil service, traditional profession or the Academy, was remoulded to open up universities to a much larger proportion of the population with a consequently greater variety of student characteristics, needs and ambitions. For this and other reasons, the university sector broadly shifted from the pursuit of knowledge as an end in itself towards more utilitarian ends. Associated with this shift were more specialised and professionalised roles within universities, greater cross-institutional competition nationally and internationally; corporate dependence and sponsorship; more standardised curriculums and efforts to provide students with practical 'employable' skills and competences.[14]

Interest in the 'role' of universities in society reflects the rise of a functionalist reading of the world in which society is a system divided into sectors and groups that each play a unique part. Still prominent today, this view of universities not only specifically encourages them to become more *useful* to society (better fitted to the whole), but fundamentally disputes the idea that universities are independent entities, free to choose what they become (e.g. on the basis of an essentialist truth or defining ideal or philosophy). Instead, a functionalist lens emphasises the relationship between universities and the world and underlines the influence of external drivers. From this perspective, the form and shape of universities is less a response to academic or educational choices (e.g. about the nature and role of pedagogy, curriculum and research) and more a reflection of contemporary culture and structures.

Through this lens, the increasing focus of universities on managerialism, efficiency and competition over the last century reflects changes in the broader environment. Even Humboldt's Enlightenment-era University of Berlin model with its strong emphasis on academic freedom, and academic purpose was located *inside,* not outside, the existing ideology and class structure system.[15]

Sociologist Joseph Ben-David argues that it was not 'the idea of the university' put forward by German philosophers such as Kant and Hegel that shaped Humboldtian universities as sites of secular learning, but rather tensions in society at the time that oscillated between innovation and rigidity. The upshot is that 'the status and the freedom of the university, seemingly so well established and secure, were as a matter of fact precarious'.[16] The same holds true now, as universities remain contested and precarious entities—perhaps even more so.

Marketisation, globalisation and massification have created shifts in both the real and perceived contributions of the role of the university. The limitations of the neoliberal university and its over emphasis on 'sponsorism' and corporatisation aligned with commercial interests have given birth to *Critical University Studies*. In the last few decades growing critiques of academic capitalism reflect increasing levels of frustration and anger around what is being 'lost, jettisoned, damaged or destroyed' within higher education.[17] This includes a marked shift away from a public model of higher education towards a privatised model that has raised serious questions about academic labour and student debt, among other issues of precarity and vulnerability in the face of instituional reform and restructure. The negative impact on academic work, student learning and the nature of institutional politics are particular points of contention within an increasingly audit-oriented culture, that too often prioritises management by metrics over the quality of student and staff experiences.

In higher education policy, pedagogy and practice in a climate of growth, this manifests as velocity over quality; project not process led planning; circumscribed community involvement as consumers/stakeholders not students or citizens; sections of higher education moving into shady wings beyond scrutiny (commercial in confidence provisions of public-private partnerships); and dubious and possibly self-serving strategic policy and planning processes and techniques especially in the area of growth modelling and budget forecasts that go to the core of the increasingly privatised model of university financing.[18]

In *A Fractured Profession*, David Johnson argues this emphasis on growth and profit in higher education is fundamentally re-shaping the purpose and contribution of the university. In particular the commercialisation of knowledge which creates a conflicting role for academics in serving the 'advancement of knowledge toward particular—and financial—ends'.[19] Similarly, Christopher Newfield highlights in *The Great Mistake: How we wrecked public universities and how we can fix them*, that the current business model for universities to improve growth and market-efficiencies has led to what he describes as 'a death spiral' for higher education. This includes a national student debt crisis, lower educational quality and an overreach of property investment and facilities.[20] Whilst in *Public Universities, Managerialism and the Value of Higher Education*, Rob Watts points to what he calls 'market-crazed governance' which has led to a situation where,

> not only the normal teacher-student relationship is inverted, academic professional autonomy is eroded, and many students are short-changed, but where universities are becoming places whose leaders are no longer prepared to tell the truth, and too few academics are prepared to insist they do.[21]

This 'new normal' is more than a crisis of universities, it is a crisis of government and society (in a Foucauldian sense)[22] which creates significant challenges, but also the space for alternative 'governmentalities'. Susan Hyatt and colleagues highlight that the close relationships that have been forged between higher education and privatised and corporate interests have been accompanied by 'new forms of governance, [that] produce new subject-positions among faculty and students and enable new approaches to teaching, curricula, research, and everyday practices'.[23] Thus, while growth remains the touchstone in higher education, there are shifting coalitions and communities of practice in higher education emerging which subvert the dominance of any one economic or political agenda—although largely still operating within the institutional and political status quo.

To better understand the progressive role of the university in society requires a critical focus on how higher education helps to shape and govern the world and vice versa. This includes the capacity to contest, alter and adapt the dominant practices and tools that have been fashioned to date—which in universities are proving to be as resilient as they are broken.

The Resilience Machine

Universities are remarkably resilient despite the many challenges they face. The capacity of the university to endure as one of society's longest standing institutions is for many a testament to its value, significance and importance.[24] Unlike most governments or private companies, the university has not just withstood centuries of changing circumstances, including severe disruptions, but also grown meteorically in global size, scope and scale during that time.[25] No longer the realm of the elite, attendance at university is now accessible for many in the community. The university is by all appearances a success story in modern history, culture and society—a resilience machine?

Jim Bohland, Simin Davoudi and Jennifer Lawrence use the term 'resilience machine' to describe the vast assemblage of policies, practices and projects around the world that are responding to myriad contemporary crises with 'resilience' initiatives[26] which, like Harvey Molotoch's urban growth machine,[27] tend to collectively and invisibly reproduce dominant political and economic systems. In particular they focus on how the concept of resilience is located within and often co-opted by, a set of dominant neoliberal mentalities. Within this context, the dependencies, relationships and underlying motivations of the groups and organisations involved strongly shape what is done in the name of resilience. More specifically, the urban growth machine analogy they use underlines how urban resilience initiatives are frequently used to exploit the economic potential of developments, manipulate the system to maximise growth and profit while convincing the public that upward growth models are important—nay necessary—for long-term security, jobs and prospects (i.e. resilience).

Universities are often among the key institutions enrolled in urban growth machines and associated resilience initiatives in different contexts, including as partners in their local regions. Universities can be seen as the *targets* of growth and resilience logics, often with the two in tandem. Reading universities as themselves examples of growth/resilience machines brings to the fore the pervasive drive within higher education to exploit the economic potential of knowledge generation and the concommitant role of credentialisation.

This also highlights continual efforts to maximise growth and profit out of *all* activities and assets including the financialisation of buildings and outsourcing of core services, as well as the normalisation of these agendas internally and externally with university staff, students and other 'stakeholders'. Given universities' knowledge creation, education and public engagement roles, and associated capacity for authoritative dissemination and normalisation of select discourses, they are well placed to advance a teleological 'desire for growth' within society. It is a discourse many then seek to profit from by positioning themselves as a vital passage point for others' individual and organisational success.

It is important to recognise that universities and cities do not have to be, and are not all, like this. Similarly, resilience, like the SDGs, does not have to be associated with perpetual growth and neoliberalism. The concept of resilience is highly ambiguous—merely a way of capturing a certain relationship to and mode of change. Depending on 'resilience for whom, what, when, where and why', it can strengthen precious elements of the world or entrench predatory ones. A major reason the world is facing such a crisis of unsustainability is that too many of the things that need to change (e.g. fossil-fuel driven car cultures, capitalist greed) are proving highly resilient, while the resilience of things that we desperately need to preserve (e.g. natural ecosystems, social systems of care) are being systematically eroded.[28]

Resilience therefore is not inherently desirable, but malleable and value-neutral: equally capable of being put to regressive *and* progressive agendas. If an entity (such as a university in the midst of COVID-19) simply adopts the goal of 'being resilient' without reflection and in the absence of other guiding principles, ethics or values, it is unlikely to change current conditions. As a result, it is likely to not only be poorly positioned to respond well to the next disturbance, but it is squandering an important opportunity for progressive real-world impact. As Bohland and colleagues note, rather than just depoliticised 'calls for more resilience' we need to inquire 'into the logics that have created the demand [for resilience] in the first place'.[29]

Although the current 'resilience machine' in cities and the universities they harness may be largely driven by the growth machine, there is the potential to use resilience building efforts to foster critique, experimentation and learning alongside—and in resistance to—an emphasis on economic resilience rhetoric.[30] This involves exploring different ways to (re)-imagine what a more just, equitable and democratic world might look like. It also involves recognising the ecological and physical context of universities. Actual 'resilience science' is about the resilience dynamics of social-ecological systems at interlinked scales, including the planet. It is partly thanks to resilience science that the unsustainability of current trajectories is now apparent.

Significantly, what is meant by resilience in this context is not the same as the dominant notion of 'bounce back' that stems from engineering. In resilience science and related ecological fields, resilience refers to the capacity to 'bounce-forward' through re-organisation and adaptation.[31] When the entity in question incorporates the social dimension, resilience often depends on social learning and citizen engagement. In contrast to the boundedness of entities assumed in engineering modes of resilience, ecological and evolutionary modes of resilience also emphasise the relative openness of targets for resilience, such as 'communities'.[32]

When applied to universities, this sort of resilience lens calls into view their multifaceted context and dynamism. In so doing it resonates with what Ron Barnett calls the 'ecological university' which we highlighted at the start of the chapter—one positioned within, conscious of, and caring towards, seven 'ecosystems' including the natural environment, but also knowledge, social institutions, persons, the economy, learning and culture. To be ecological, universities need to not just *sustain* themselves or acknowledge they are *from the world*—but be *for the world*—to help it change for the better. To achieve this, many universities, and the higher education sector in general, needs to question how they themselves might need to change. As he explains:

> To pick up just one ecosystem, the knowledge ecology, its sustainability [as in persistence] is not at issue. Rather, the issue is one of its ever-fuller flourishing: does it exhibit a due diversity with, say, non-scientific forms of thought being accorded legitimacy? Is there an ever-greater circulation of ideas in a polity? Is there a continuing creativity in those ideas? Do the

dominant knowledge frameworks include those of peoples across the world, including the South (Connell, 2007) and indigenous traditions (de Sousa Santos, 2016)? Does the knowledge in a society reach out so as to form ever-wider publics? Is there a healthy degree of continuing scepticism, debate and even rivalry, with groups pitted against each other in critical dialogue? If the answer to questions such as these is 'yes', then we are not in the presence of a knowledge ecology that is being sustained, but rather one that is being strengthened and developed.

Such an ecology is the sort that an 'ecological university' needs to help generate. In other words, Barnett's notion of ecology is aligned with ecological and evolutionary resilience, rather than the engineering or bounce back resilience that is instead what he expresses as 'sustainability'.

The take home message here is that we in the higher education sector need to be careful about what is sustained and what is changed. This is especially the case given the highly mixed character of universities, full of desirable and undesirable elements. As Barnett continues, even if an ecosystem (e.g. a knowledge ecosystem) that a university is part of is flourishing, we cannot take for granted that the presence of the university and its attendant networks such as academic publishers are wholly helpful in generating the positive outcome, and are not in fact a hindrance in at least some ways.

To return to the resilience of the university as an institution, the highly mixed character of its contemporary form—'the modern university'—means that its resilience is a double-edged sword. Resilience thinking and action *can* be mobilised for positive institutional and societal change (e.g. reducing the precarity of casualised workers across sectors), but can equally be put to use to perpetuate political and economic power and the status quo. However, this requires that the *de-politicisation* of ideas around resilience must therefore be countered by the *re-politicisation* of resilience and re-imagined as a more transformative and regenerative agenda. The hope for a more sustainable future.

We argue that the 'resilience machine' that is sustaining the university as an institution must be brought to light and examined through critical questions about the values and impact of the university. Such questions include:

- What is the university seeking to make resilient, from what and how? What types of resilience are being pursued and why?
- What values are inscribed and prescribed in universities in the name of resilience? What social and political effects are resilience initiatives encouraging and why? What role is self-interest playing?
- What creative and critical potential exists in alternative resilience discourses, policies and practices? How might we imagine it?[33]

Whether the future higher education enterprise proves to be a tool for economic expansion or a space that supports social equity and planetary justice depends on which elements of the modern university, including which values and goals, prove to be most resilient. As indicated above, this is not just a matter of choice for universities—they are strongly shaped by their external context as well. Shifting community expectations about what higher education can and should be are therefore of crucial importance.

Universities of Utopia/Dystopia

The 500th anniversary of Thomas More's novel *Utopia* in 2016 sparked a surge of questions about the ideal university as opposed to the idea of the university (although the two can converge). Moore's notion of Utopia translates as 'non-place' and/or 'good place'. It describes a desired future place or way of being. In *Utopia* the desired place was a fair society, as More described, 'I can perceave nothing but a certein conspiracy if rich men procuring their own commodities under the name and title of commonwealth'.[34] This has resonance with the discussions above around the neoliberal agenda driving universities as 'resilience machines'. As Terry Eagleton argues, 'one of Utopia's most striking aspects is its contemporaneity—the way in which the greedy, unscrupulous and useless are just as much in evidence now as in 1516'.[35]

These satirical themes are also reflected in the Australian comedy television series 'Utopia' (Dreamland) which follows the fortunes of a newly set up government agency focused on the delivery of major projects.[36] In microscopic detail the series sends up the collision between grandiose plans, political self-interest and self-promotion, institutional white elephant projects, bureaucratic bungling and the mundanity of everyday activities. Some have argued that the same tragi-comical combination of self-importance and incompetence characterises modern universities. Mark Gatenby, for example, describes the contemporary university as characterised by 'burdens of meddling, managerialism, bureaucracy and consumerism'. Significantly he asks: if higher education is becoming a corrupted, capitalist artefact, how can we live well with the universities of the future?[37]

Perhaps the aspect of universities most likely to determine how close the actual university is to the ideal is how it is funded. The public university model, which has been the dominant mode of higher education since World War Two and much of the twentieth century, has historically been funded largely by the state and student tuition fees. More recently in many countries around the globe, governments have cut back funding for universities, forcing the higher education sector to become more entrepreneurial by finding other ways to make up the funding shortfall through a business approach. This includes increasing domestic and international student numbers and raising tuition fees, philanthropy, grants and contracts, endowment, property ownership and development, and income-generating investment portfolios.[38] As the American Academy of Arts and Sciences puts it:

> As state appropriations for higher education diminish, public universities increasingly rely on other sources to advance their mission and maintain the quality of education and training they provide: tuition, philanthropy, auxiliary services, grants and contracts, and endowment and investment income. The extent to which individual public research universities rely on diverse sources of funding varies greatly by location, demographics of students served, state aid programs, and relationships with regional business and industry. Some institutions fare better than others due to generous state funding, robust philanthropic enterprises, or lucrative partnerships with local corporations.[39]

Each approach to generating alternative income has its pros and cons. Some universities, for example, are building new research centre partnerships with private sponsors with the concomitant aims of generating funds for the university and increasing opportunities for real-world engagement and impact. However, such initiatives can often benefit only a portion of staff and can generate relationships and goals that are misaligned with other university objectives. Other universities are trying to reach a wider domestic and international market using online programmes, courses and certifications delivered and undertaken at a fraction of the cost of face-to-face teaching, but raising questions about the quality of the educational experience, as we discuss further in Chap. 6.

Many universities have increased their proportion of full fee-paying international students to increase funding revenue and develop a more diverse, global student body. However, as the COVID-19 crisis has demonstrated, an (over)-reliance on international students can exacerbate the economic precarity of universities if it is disrupted or lost to competition or complacency. Taking on responsibility for large numbers of international students also means guaranteeing them appropriate levels of support and care. If this, rather than the flow of students, breaks down, universities are no less at risk, with complaints that the international student trade industry is simply viewed by universities as a 'cash cow' sullying institutions' reputations and risking a serious breach of public confidence and trust in the whole sector.[40]

Overall, in the competitive global education market, individual universities compete for income on an increasingly uneven playing field constrained by socio-spatial and economic factors such as size, reputation, postcode and location, history and accumulated debt, access to industry partners, and the state and maintenance costs of existing and future requirements for university infrastructure. Different funding choices are available to, and variably effect, different types of universities. Across the board, many have been pushed to become more 'entrepreneurial' in their internal operations. Many have taken aggressive measures to reduce operational costs by cutting faculty and staff positions, reducing tenure to contract or casual positions, eliminating or streamlining course offerings, outsourcing core services and operations and instigating performance metrics focused on success in external funding, in efforts to increase institutional accountability and efficiencies.

At the same time, there is growing pressure on universities to better justify their societal role and public support in the face of societies 'grand challenge' such as the need for action on climate change and the SDGs. The new dominance of an economic logic at work in both the strategic and operational levels of universities is generating increasing concern that the university has become almost unrecognisable to its older ideal. As Simon Marginson puts it in his personal reflections on the relationship between higher education and the common good,

> I have been compelled by the questionable foundations, gaps and internal tensions in the standard thinking about higher education and its social and economic roles … the faith that the installation of competitive markets into higher education will lead to better quality and greater responsiveness to the needs of students would be touching in its naivety if it wasn't also so destructive. The inability of economics to adjust for the particular character of social production in the higher education sector … continues to do much damage.[41]

Such damage includes environmental damage. It is increasingly clear that existing approaches to universities are not adequately meeting societal and planetary needs. Nor are they meeting societal expectations or building public trust. If academic institutions are to secure their future, they need to demonstrate a genuine commitment and capacity to work with others to achieve the transformational changes needed. As Barnett argues in *The Ecological University* this is a feasible utopia: one which recognises that the university is 'a story without an end' and that 'new opportunities may be opening up for universities to engage in, and even enlarge, the public realm by forging new relationships with the world/earth'.[42] Part of this challenge—and opportunity—is to re-imagine the nature of the relationship between the SDGs and higher education as part of a broader social contract focused on a sustainable future.

Seeking the Good University

As visions of alternative universes, utopian thinking offers a device for simultaneously disrupting or unsettling the complacency of the present, as a way of projecting the hopes and dreams that drive action in different future-oriented directions. The early thinking around the notion of utopia is linked to Plato's Republic and Laws which hold that the ultimate end of the system is to bring about the greatest possible happiness in the city. The idea of 'the collective good' dominates all aspects of Plato's utopian communitarian world and is synonymous with quality and truth.[43]

In her book '*The Good University—What Universities actually do and why it's time for a radical change*', feminist sociologist Raewyn Connell makes the case that whilst fragments of the good university already exist, choices must be made as to what types of higher education futures are desirable. 'There are better futures we can choose for universities by collective choice, not the individual decision of a market consumer.'[44] The good university and good university systems should be collective and cooperative, operating at the level of society: 'it has to be made and re-made, daily and from generation to generation to make the commitment and struggle worthwhile'.[45] To this end, Connell outlines five principles for a 'Good University':

- *Democratic*—develops a democratic culture, operates in a democratic way, and serves a democratic purpose for society.
- *Engaged*—is fully present for society, responsive to societal needs at the local and global scale.
- *Truthful*—in detailing university operations and in how it presents itself to fulfil its purpose to serve society.
- *Creative*—by embracing the dynamism of knowledge formation and educational processes, expanding, devising, imagining, patterning and linking the different forms.
- *Sustainable*—the capacity to flourish over time, creating conditions of renewal and resilience in the face of disruption, change and political pressure, and responsible use of resources, similar to the imagined relation between Ron Barnett's ecological university and the natural environment.

Like the question of university resilience, the quest for the good university raises a number of critical questions—Good for whom? Good when? Good for what? Who gets to decide what is good or even good enough? Sceptics and critics point to the chimerical quality of utopian vision, which is dependent on ideology, culture, politics and perspective. The version a given university supports will depend on what the institution seeks to see flourish at a given point in time, reflecting its history and contemporary context. No blueprints for a good university exist, and even if they did, they would not guarantee outcomes. As Ash Amin notes, 'The concept of good does not track unmodified across space and time'.[46]

Contemporary contestations over the 'idea of the university' are made manifest through a series of unresolved and relational tensions. These play out on a spectrum across the different university models, sectors and roles. Education policy historian Robert Anderson argues that key tensions include those between: the goals of teaching and research; academic autonomy and corporate accountability; scholarly learning for its own sake and the achievement of qualifications and skills; transmitting established knowledge and challenging the status quo; the connection of universities to the state and private sector and the need to maintain critical distance; the reproduction of existing power structures and renewal from below through resistance and/or social mobility; commitment to an international community and a national identity; and serving the economy whilst addressing transformative individual and societal change.[47] Raewyn Connell highlights:

> There's an angry, sometimes anguished debate inside universities. Critics speak of outdated pedagogy, exploitation of young staff, distorted and even faked research, outrageous fees, outrageous pay for top managers, corporate rip-offs, corruption, sexism, racism and mickey-mouse degrees ... there is criticism from outside the university too ... contemptuous of university educated 'elites' and university-based science.[48]

Given this bi-directional critique, Connell argues that we need to rethink and debate the fundamentals of what universities do. Her version of a good university is one driven by social good rather than profit—but for others it may involve wealth accumulation and elite prestige. Still others may prioritise a decolonised university that is respectful, inclusive and fair, while global access and technological sophistication may represent the 'good university' for some. The answer may be a combination of these or some other vision altogether. Importantly, a key question is are universities 'good enough'—not for what the world wants, but for what the world needs?

Good Enough Universities?

Are our universities good enough to face the twenty-first-century challenges? How would we know? The contemporary university is no longer cloistered away in the tower of academe but inescapably embedded in the world. Universities are committed to a mission that underpins their purpose and function in society as centres of new knowledge, understanding, skills and experience, through research, learning and teaching, leadership, outreach and service to society. As proponents of progress, choice, debate and engagement, universities help 'set the pace for humanity' in the key areas of society, culture, economics and the environment.

But the role of higher education 'will not trigger the development of a more egalitarian society on its own'.[49] Being out in the world, universities have many masters: national governments are one, but also local and regional/state governments, industry partners and other private funders, and not-for-profits, as well as local and broader community. Combined with internal masters such as discipline-specific peer-reviewers, associations, publishers, and of course students armed with evaluations, universities' capacity to initiate progressive and meaningful change is shaped by many groups. Thus:

> assumptions about higher education being able to independently and single-handedly effect the betterment of society (in tackling inequality or in stimulating innovation of the economy) are proven to be strongly exaggerated. Rather, higher education can be one of the critical factors—a tool—affecting these processes, but whether it will be used for this purpose remains a question of political choice.[50]

Universities cannot affect change on their own, but they can be effective when working with others. Moreover, their role is not just one among many, but a special one. Whatever their size, shape, scale or funding model, universities have a unique capacity to cultivate and share ideas, methods and frameworks for the betterment of society. They can act as agents of change in multiple ways, such as by creating 'critical space' and engaging in 'dialogue, debate, and development of proposals and programs for social change, with the ultimate objective of engagement in the public sphere', and by designing and making improved artefacts for use in the world.[51]

The 'idea of the modern university' is a complex entanglement of what Hannah Arendt described in 'The Human Condition' as the two images of human activity.[52] The first is *Animal laboran* whom she critiques as becoming so absorbed in their tasks that they get lost in the act of making and doing such that work becomes an end into itself. The second is *Homo faber* who she favourably notes is focused on critical thinking, judgement and making a life in common. Whilst *Animal laboran* is fixated on the question of 'how', *Homo faber* asks 'why'. For Arendt, society is afflicted with an overly active desire to 'do' without considering for what purpose. Thoughtfulness-in-action, she argues, is the critical and necessary human response to the world, especially during dark times.

Arendt's thesis is that history has shown that the capacity of humans to build, make, do, manage, organise, invent or innovate is not in itself enough. Her faith was in critical speech, action, politics and reflection—uniquely human capacities that she believed will save humanity from itself. A life without critical reflective speech and action, she argued, 'is literally dead to the world'. She refers to a quote by physicist Robert Oppenheimer who invented the atomic bomb: 'you see something that is technically sweet, and you go ahead and do it, and you argue about what to do about it only after you have had your success'.[53]

Richard Sennett, a former student of Arendt, similarly advocates for critical thought but argues that Arendt over-emphasised the divide between 'doing' and 'thinking'. Rather than choosing between them, he argued, we must always ask 'why' as well as 'how'. To understand *Homo faber's* role, he suggests:

> we have to conceive of the dignity of labour differently. ... *Homo faber* acquires honour by practicing in a way whose terms are modest and this

ethic of making modestly implies in turn a certain relationship with how we dwell and who we are.[54]

Sennett stresses that his emphasis on the dignity of labour—on *Animal laborans*—is not to be confused with a romantic endorsement of craft-building in and of itself. There is nothing inevitably ethical about craft-building, he argued. The craftsperson's desire for quality, for example, can still pose a danger if obsession with the task deforms the work itself and allows it to become morally ambiguous.[55] Or the craftsperson's vision can get lost, as Ash Amin describes of modern cities, and by extension our city universities:

> as shadowlands: anonymous, homogenous and lacking character and identity; endless unhealthy, tiring, overwhelming, confusing, alienating; with little connectivity or potency as demos—the populace of a democracy as a political unit.[56]

For Sennett, what we ultimately need is an integration of *Animal laborans* and *Homo faber* in the form of active citizenship—an approach to being and doing that seeks to find expression in the world through thinking and feeling, action and reflection, problem solving and problem finding as a co-constituted rhythm, not separate activities.

For universities, active citizenship means not getting lost in the business of doing and surviving and making, nor offering mere reflection and judgement. It is about active but critical engagement with the world all universities are part of. Today an unavoidable consequence of such engagement is awareness of the many problems the world is facing—including those covered by the SDGs. Only by working with others to tackle these problems in insightful and practical ways can a university be considered 'good enough' for the contemporary context.

The Urgent Need for Maintenance, Repair and Regeneration

Besides a proliferation of problems, the contemporary context that universities need to respond to is characterised by a need for deeper ambition. Not only minimal solutions or short-term preventative action is needed but so too is repair, regeneration and maintenance. Sustainable development today needs to involve not only the minimisation of negatives such as pollution or illness, but the active generation of positives such as ecosystem health and human wellbeing in order to redress the immense amount of damage already done, improve the capacity to cope with future stresses, and protect what we value and care about into the future.

Ash Amin offers four registers of a care-ethic that begin to point to what such sustainable development requires: repair, relatedness, rights and re-enchantment. We can use these registers to help progress our understanding of what the Good University might look like. Through a focus on a politics of *Repair* the good university commits to accessible and affordable infrastructure expressed through practices of care and solidarity. The emphasis on *Relatedness* orients the good university towards an ethos that is socially and environmentally just with a strong obligation 'towards the insider and the outsider' regardless of ethnicity, race, gender, age or ability. Amin's focus on *Rights* is the citizen's right to the university for the many, not just the few and the creation of an 'open' civic culture that works democratically with difference and disagreement. The final register is a politics of *Re-enchantment* through a focus on civic-mindedness as a counterpoint to commodification and homogenisation on the one hand, and disinterested and disengaged individualism on the other.[57]

Many of the grand challenges we currently face stem from poor maintenance of the systems we have created or rely on. Doing things cheaply is a natural outcome of the short-term focus that characterises commodification, capitalism and political cycles—and this extends to the functioning of universities. As Jason Moore and Raj Patel argue in their book *A History of the World in Seven Cheap Things*, '*Cheap* is the opposite of a bargain—*cheapening* is a set of strategies to control a wider web of life'.[58] We are all now paying the price of not maintaining healthy landscapes, waterways and cultural and social relationships. One of the fundamental changes required is to re-value the care work that maintenance and reproduction rely on, as Amin indicates and feminist scholars have long argued, given the way that the symbolic and demographic feminisation of care is entwined with its social and economic devaluation.

As Moore and Patel discuss, one of the things that has been cheapened by capitalism is care itself, even as the commodification of care has made it financially expensive for many to receive. The drivers and consequences of this are evident in the university system where production is valued over re-production (maintaining the enabling conditions) within research activities and in the status attributed to research over teaching. They are also evident within the SDG agenda, which includes both implicit calls for care work—care of people, communities, institutions, settlements, ecosystems and the climate—and endorsement of economic growth and development.

Beyond maintenance, the degraded and endangered state of the world demands serious investment in repair and regeneration. Valuing repair work and associated objectives such as retrofitting are key to curbing excessive consumption and production of the sort that SDG 8 demands we rethink. They are also key to redressing the environmental injustices that continually generate harm and constrain positive developments in cheapened landscapes and communities around the world. The profitability of the extractive industries, for example, continues to rely on their ability to walk away from damage—leaving behind what Val Plumwood referred to as 'shadow places'.[59] Making visible these injustices and the pressing need for repair and restoration has to be part of the mandate of sustainable development in the Anthropocene.

Beyond repair and restoration, regeneration is also needed. Like many others increasingly grasping the positive potential of this idea, we understand regeneration to mean more than a neutralisation of negatives or a minimalist attitude of compliance. Regeneration is about nurturing new life and potential itself. It is about reclaiming the development ideal of a better possible world and helping cultivate it in a genuine, care-full, life-affirming way, one that is necessarily experimental and courageous in facing up to failures and trying again—living and working with the trouble of our times.

The End of the University as We Know It?

Alternatives to the dominant model of neoliberalised universities exist, and they offer insights into what a re-imagined university sector may look like. One approach is 'civic universities' that reframe cities as inherently dynamic public places full of potential, active learning and innovation, making universities just one node among many. This blurring of the boundary between the university and civil society is not just about a greater extension of the university out into society—as the idea of adding on public engagement to universities as a peripheral 'Third Mission' alongside Research and Learning and Teaching suggests. Rather, it is about using the university as a common, as a platform for civil society to co-produce place and common goods. John Goddard characterises civic universities as anchor institutions, which are 'not just in the place but of the place'. He offers seven characteristics that distinguish the civic university:

1. It is actively engaged with the wider world as well as the local community of the place in which it is located.
2. It takes a holistic approach to engagement, seeing it as institution-wide activity and not confined to specific individuals or teams.
3. It has a strong sense of place— it recognises the extent to which its location helps to form its unique identity as an institution.
4. It has a sense of purpose— understanding not just what it is good at, but what it is good for.
5. It is willing to invest in order to have impact beyond the academy.
6. It is transparent and accountable to its stakeholders and the wider public.
7. It uses innovative methodologies and team building in its engagement activities with the world at large.[60]

All of these characteristics are ones that could foster genuine engagement with the SDG agenda and facilitate its effective localisation. In contrast to an approach that cordons SDG work off into certain courses, projects or outreach initiatives, or eschews practical action, the civic university represents the sort of approach to higher education that the pressing challenges of the SDGs demand. Not all universities can become *locally* oriented civic universities per se, given the diversity of contexts in which they exist. Nor is it necessary, given that the SDGs still require a strong international, national and regional orientation and level of cooperation, and that 'community' today has many forms including distanced virtual ones.[61] But the ethos of the civic university holds lessons for all universities, in particular its commitment to engagement, purpose and impact (Fig. 3.2).

A second thought-provoking alternative to the dominant model of universities is the 'free university', which directly addresses and reframes neoliberal modes of development in higher education. The vision of 'free universities as commons' builds on a rich tradition of feminist, anti-racist and working-class struggles in the development of postcapitalist imaginaries in academia.[62] As Esra Erdem describes, this is about universities as 'grassroots spaces created by a community for the sharing of knowledge in which knowledge and ideas can be freely shared among equals'. In such institutions, 'space is not given: it has to be established and occupied'.[63] Erdem highlights four key principles/themes inspired by the community

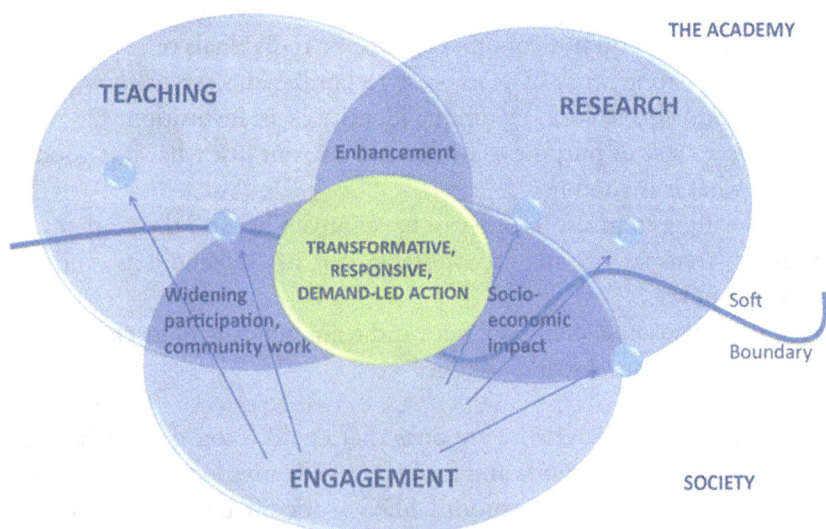

Fig. 3.2 The civic university. (From Goddard 2018, p. 263)

economies agenda for fundamentally re-imagining the academy as community commons:

- *Access*: The principle that higher education should be socially inclusive and foster the sharing of knowledge and specifically targeting restrictions in three key areas: university admission criteria; tuition fees; and intellectual property rights including the sharing and access to knowledge. This involves making resources available to the community and re-shaping spaces of learning to include community settings such as parks, libraries, churches, trade union halls, community centres, cafés, bookstores, galleries. Access also involves reducing economic barriers to higher education and challenging universities as banks of commodified knowledge or 'teaching factories',[64] committing instead to inclusive, collaborative learning.[65]
- *Commoning practices:* Grounded in the critical pedagogy of Paulo Freire, commoning practices are the social labour undertaken by a community to produce and sustain collective resources. In universities, this is about acknowledging the diversity of community experiences and knowledges and the multiplicity of skills involved in everyday work (including that involved in administrative, coordinating and logistical work). It is about acknowledging the potential for collective labour to not simply replicate the existing hierarchies of mainstream academia (e.g. gender, academic rank etc.) but to consciously create commons and solidarity.[66]
- *Collective self-management:* The emphasis here is on processes of collective decision-making that critique hierarchical university structures and enable more participatory forms of decision-making. Informality, autonomy and responsibility for the academic commons are key shared characteristics across diverse decision-making practices.[67]
- *Community*: This final principle is about the development of alternative power and knowledge relations that help rebuild a sense of community as part of an academic commons. Community-building through the free university includes nurturing a sense of learning, belonging and commitment, 'being-in-common'.[68] This includes values and ethical principles such as equality, reciprocity, trust, localness, social justice and freedom, where the latter is understood as the right

to partake in education and a shift from market exchange to gift economies in higher education.[69]

There are many international examples of the 'free university' in practice including *Universidad de la Tierra* (Unitierra) in Mexico for Indigenous and poor urban communities,[70] the *People's Free University* in Canada,[71] the *Universidad Trashumante* in Argentina, the *Social Science Centre* in the UK and *Solidarity Academies* in Turkey[72] to name a few. Whilst diverse in context, ambitions and practice their shared agenda as part of the free university movement is a profound resistance to, and critique of, the neoliberal development model of higher education (e.g. the commodification of education, the re-organisation of the labour process, the enclosure of knowledge and the financialisation of student debt).[73] Collectively they demonstrate alternatives to how learning and teaching can be organised around the principles of commons,

> shared by a community of users/producers, who also define the modes of use and production, distribution and circulation of these resources through democratic and horizontal forms of governance.[74]

In addition to the civic university and free university models there is a range of other alternative forms, which collectively call into question the naturalisation of the dominant modern university model. For example, in *The Good University*, Raewyn Connell highlights practical examples of alternative university manifestos that offer principles, pedagogies and processes for transformative change.[75] The *Slow University*, for example, seeks to subvert the corporatisation culture and 'speed-up' of universities in the quest for efficiency.[76] As part of the diverse slow movement (e.g. slow food), the Slow University seeks to advance alternative/unorthodox approaches that enable community-based initiatives, sustainability and social equity in the face of 'fast capitalism'. The slow movement is an alternative development narrative deployed through a diverse coalition of actors. In contrast to the mainstream development agenda characterised by homogenisation, standardisation, corporatisation, insensitivity to local history and culture, and conditions of inequity, the slow agenda is characterised by grassroots activity, sensitivity to local history/culture and specific attention to issues of sustainability and equity.[77]

Others are turning to alternative modes of development to re-imagine a form of higher education that pushes 'beyond the limitations of the paradigm of modernity and the neoliberal ideology that commits us inescapably to economic growth at the expense of the environment'.[78] As Eleanor Brown and Tristan McCowan argue:

> If we are committed to the idea of sustainability, which has not been well served by these ideologies and agendas, we might want to consider what 'sustainable development' looks like from worldviews with a quite different ontology from the European modernity upon which our development discourse has been based.[79]

They highlight *Buen vivir* in the Latin American context and *Ubuntu* in the African context as rich Indigenous worldviews with explicit lessons for modern higher education. Drawing on such insights, they offer broad principles for an education model designed to cultivate the conditions for a sustainable future:

- *Epistemological pluralism*: acknowledging and transiting between different forms of knowing.
- *Porosity of boundaries:* non-rigid classification of the educational space, education professionals and disciplines.
- *Holism of learning*: bringing together of the manual, practical, technical, abstract, aesthetic and spiritual.
- *Cooperativism*: avoidance of competition-based education and the consequent progressive filtering out of students from level to level.
- *Compassion and nonviolence*: recognition of the importance of peace in all aspects of life, including nonviolent communication.
- *Collectivism*: learning collectively within a web of relationships between people and with the non-human world.
- *Meaningful livelihoods*: a link with enriching forms of work (rather than alienating employability).
- *Living the present*: education as a state of being, not aimed at the exchange value of qualifications.[80]

The *Campaign for a Global Curriculum of Social Solidarity Economy* for example promotes the construction of 'another possible education and economy' through connections between social actors and social movements. An educational initiative developed by a collective of organisations in the international Education and Social and Solidarity Economy Network (REESS), the campaign calls for a Global Curriculum of Social Solidarity Economy, by which it means:

> the plans of studies, educational proposals, knowledge, epistemologies, methodologies, science and practice of formal education non-formal and informal, developed around the world, in schools, universities, social movements, cooperatives, trade unions, associations, indigenous peasant communities and Afro-descendants, while building a just, sustainable not capitalist economy.[81]

Significantly, the REESS campaign is explicit about the value of engaging with the SDGs as an ambitious and transformative change-based agenda, and about the value of its alternative solidarity economy to the SDGs. As the Network puts it:

> The basis for the achievement of all the Sustainable Development Goals ... stimulates criticism of the current model of predatory economy and unsustainable patterns of production and consumption but also promotes and strengthens actions that represent alternatives to that model. At the same time, inclusive, equitable and quality education that promotes lifelong learning opportunities will actually exist if we strengthen the idea that another economy is possible.[82]

Many more alternative, progressive approaches to universities already exist or are emerging, suggesting that more people may be committed to the idea and ideal of the university than its specific, dominant modern form. The existence of such alternatives is an important reminder to look beyond the often uninspiring and concerning characteristics of today's universities to remember and reimagine what they can be. When we do so, it is more apparent than ever that universities are vital to progressing the SDG agenda (as a diverse range of organisations in their own right and as enablers of others) and the SDGs are vital to reshaping universities.

In their best light, the SDGs offer more than a topic for research or teaching or a competitive global indicator for the higher education sector. They can offer sustenance in everyday struggles and opportunities to subvert established processes, and a lens through which to analyse, critique, adapt and improve the myriad development processes universities are enmeshed in. For such institutions, this requires not simply mapping existing SDGs capabilities, but committing and delivering ethical SDG-informed innovation at all scales and building the SDGs into their ethos and institutional architecture.

The Age of Reciprocity and Change

As we have pointed out, universities are not isolated ivory towers, floating free from the rest of the world. They are of the world, as their remaking as corporations over the last fifty years illustrates. For better or worse, they are being constantly reshaped by the world and, for better or worse, they are continually shaping the world in ways that far exceed laborious efforts at 'engagement'. The phrase 'for better or worse' is the key. It begs the question: Will those in the diverse communities that constitute and influence universities strive to the institutions and their outcomes with care, or will they act as if the mutual shaping is not happening and accept the consequences?

We believe the reciprocal relationship between the SDGs and universities can help chart a careful path between the dead ends of disconnection and false connection. Disconnection is about perpetuating the myth of the ivory tower by presuming they are unchangeable, untouchable or innocent. As Isaac Kamola argues,

> Despite being located within vast overdetermined social relationships, those students, scholars, and administrators inhabiting the world of higher education often imagine universities as extra-worldly spaces from which to orbit—and gaze down upon—the world below. In claiming to simply reflect upon the world, seeing it as it actually is, the university often fades

from the foreground, cropped out of the imaginary. In this process, colleges and universities increasingly are perceived as ivory towers located above and outside the world.[83]

This conception of universities as 'extra-worldly' 'ivory towers' is of course far from new, and in some senses is almost as old as the idea of the university itself. But it is one that is increasingly hard to sustain. As Kamola continues,

> In reality, however, there is no outside from which to view the world as a single thing, global or otherwise. A university is not a capsule floating outside the world's orbit. As such, academic knowledge is never merely a snapshot of the world outside itself.[84]

Many feminist, Indigenous and postcolonial scholars have criticised the idea that knowledge can emerge from a 'view from nowhere' and that universities can deny their position within and obligations to engage with 'the real world'. John Brennan and Allen Cochrane similarly argue that the idea of universities as universal and place-less 'is no longer a helpful starting point'.[85] It is now recognised and expected that universities are 'of the world' and part of society, albeit still a special part.

Just how universities are imagined to be of the world, though, varies widely, as the discussion throughout this chapter illustrates. One of the risks is that universities seek to establish strong connections 'with the world' but that such connections are *false*, either in the sense of not grasping essential realities of the world such as the severity of climate change, or not being genuine connections and being, for example, for show only. Arguably the dominant style of the connected, 'worldly' university today is one that maintains a focus on the global/universal scale but focuses only on aspects of it, namely the global economy, city networks and elite institutions. The ideal is of a 'world class university' that combines claims to extra-worldly universal knowledge and research excellence on the one hand, with claims to global economic, urban and institutional connectedness on the other. The world class university is imagined as a privileged node in global circulations of resources, people and knowledge, not shut off from the world but confidently leading it from on high.

Many modern universities are also or instead emplaced within more specific innovation systems, districts, precincts, clusters and other 'triple helix' initiatives to help drive economic development from the city scale to the globe. This now-dominant Americanised model of the university 'blurs the boundaries between public/private, and non-profit/for-profit' and 'emphasises university entrepreneurship and university–industry partnerships'. In doing so, 'notions of higher education as a producer of public goods and a cultural project are marginalised'.[86] As Simon Marginson concludes:

> There is no reciprocity here. The University is accountable to capital, but capital is not accountable to the University or subordinated to its logics of teaching/learning and knowledge exchange.[87]

The worldliness of this capitalist university is not one in which the university reaches out into the world, as much as one in which capitalism reaches into it. Although the related knowledge economy discourse 'talks of universities' potential to transform societies', it 'may be limiting this potential if one values societal transformation in all its diverse, non-economistic dimensions'.[88]

It is in the context of these various versions of the global university— the place-less universal university of old, the elitist 'world class university' and the capitalist university of innovation systems, that the global orientation of the SDGs is situated. Some critics within universities reject the SDGs because its global character is seen to perpetuate the hubris and harms of the global university. However, the disconnection and false connections that characterise the typical global university are not the sort the SDGs encourage or require. Instead, they press universities to acknowledge in a far more holistic and clear-eyed way the world and planet that all institutions are part of, and to work in more genuine and effective ways to shape it for the better of all.

As we have outlined, a key theme of this book is that universities are not just enablers of change in the SDG agenda but also important *targets for* the sorts of changes it calls for particularly given that universities' diverse functions and responsibilities have far-reaching implications for the success of the SDG agenda. If the SDGs are simply a perpetuation of a smooth, unreflective global imaginary of the sort colonialists, capitalists and universities have long encouraged, it is difficult to argue that it is what the world needs. Although there is clearly a risk of this, we believe that the SDG agenda presents the opportunity to challenge this image of the world and the image of universities as conveniently disconnected from or hopelessly compromised within it. Like alternative forms of universities, the SDGs help underline that the world, including the planet, is far more-than-economic and is instead saturated with life, diversity, meaning and inhuman forces.

For universities to perform their unique function as enablers of change, they need to embrace their role as targets for change and ensure they are role modelling the sort of approaches and impacts they want to engender. The SDGs push us to consider the global scale, but it is not the disembodied space of the ivory tower myth or the ruthless machine of the global economy. Rather the SDGs provide an opportunity to simultaneously address some of the harms of a neoliberal mindset that pits individuals against individuals, departments against departments, universities against universities, nations against nations, and human growth and development against the ecological health and sustainability of earths' planetary boundaries.

Notes

1. See 'Post-pandemic futures: what does covid-19 mean for HE?, *Times Higher Education*, 7th May, accessed via (THE) newsletters@timeshighereducation.com
2. Kamola, I (2016) Situating the "global university" in South Africa, in Chou, M, Kamola, I and Pietsch, T (eds.) *The Transnational politics of higher education: Contesting the global/transforming the local*, Routledge, New York, p. 42–63.
3. Langin, K. (2021). "Pandemic hit academic mothers especially hard, new data confirm." *Science* magazine https://www.sciencemag.org/careers/2021/02/pandemic-hit-academic-mothers-especially-hard-new-data-confirm

4. See for example https://sdg.iisd.org/news/universities-declare-climate-emergency-ahead-of-climate-action-summit/ On the SDG Accord, see https://www.sdgaccord.org/
5. Zwar, C. and Lancaster, S. (2020) How can universities respond strategically to the climate emergency? *Wonk HE*, 10/2/20 https://wonkhe.com/blogs/how-universities-can-respond-strategically-to-the-climate-emergency/
6. Wynes, S., S. D. Donner, S. Tannason and N. Nabors (2019). "Academic air travel has a limited influence on professional success." *Journal of Cleaner Production* 226: 959–967. See also Shields, R. (2019). "The sustainability of international higher education: Student mobility and global climate change." *Journal of Cleaner Production* 217: 594–602. Higham, J. and X. Font (2020). "Decarbonising academia: confronting our climate hypocrisy." *Journal of Sustainable Tourism* 28(1): 1–9.
7. Barnett, R (2018). *The Ecological University: A Feasible Utopia*, London/New York, Routledge.
8. Ibid., p. 4.
9. Newman, J (2020) *The idea of the university defined and illustrated*, Glasgow, Good Press.
10. Rothblatt, S (1997) *The Modern University and its Discontents: the Fate of Newman's Legacies in Britain and America*, Cambridge, University Press. Scott, S (1993) The idea of the university in the 21st century: a British perspective, *British Journal of Educational Studies*, 41, p. 4–25.
11. Newman, J (2020).
12. DeBoick, S 2010 Newman suggest the universities soul is the mark it leaves on students, in *The Guardian*, 20th October, accessed on https://www.theguardian.com/commentisfree/2010/oct/20/john-henry-newman-idea-university-soul
13. Nybom, T (2003) The Humboldt legacy: reflections on the past, present, and future of the European university, *Higher Education Policy*, 16, p. 141–59.
14. Anderson, R (2010) *The 'idea of the University' today, History and Policy*, accessed on http://www.historyandpolicy.org/policy-papers/papers/the-idea-of-a-university-today
15. Ben-David, J and Zloczower, A (1961) The idea of the university and the academic marketplace, *European Journal of Sociology*, 2(2), p. 303–314.
16. Ibid.

17. see Palgrave Critical University Studies series—https://www.springer.com/series/14707
18. Gleeson, B and Steele, W (2010) *A Climate for Growth*, Brisbane, University of Queensland Press.
19. Johnson, D (2017) *A Fractured profession: Commercialism and conflict in academic science*, Baltimore, John Hopkins University Press.
20. Newfield, C (2016) *The great mistake: How we wrecked public universities and how we can fix them*, Baltimore, John Hopkins University Press.
21. Watts, R. (2017) *Public Universities, Managerialism and the Value of Higher Education*, London, Palgrave.
22. Foucault, M. (1991) 'Governmentality' trans. Rosi Braidotti and revised by Colin Gordon, in Burchell, G, Gordon, C and Miller, P (eds.), *The Foucault Effect: Studies in Governmentality, pp. 87–104*. Chicago, IL: University of Chicago Press.
23. Hyatt, S, Shear, B and Wright, S (Eds.) (2015) *Learning Under Neoliberalism: Ethnographies of Governance in Higher Education*, Berghahn Press.
24. Kittleson, J and Transue, P (1984) *Rebirth, reform and resilience: Universities in transition*, Columbus, Ohio Uni Press.
25. Karlson, J (2013) Resilient Universities: *Confronting Changes in a Challenging World*, Peter Lang Gmbh, Internationaler Verlag Der Wissenschaften.
26. Bohland, J, Davoudi, S, Lawrence, J (Eds) (2019) *The Resilience Machine*, New York, Routledge.
27. Molotoch, H (1976) The City as a Growth Machine: Toward a Political Economy of Place, *American Journal of Sociology*, 82(2), p. 309–322.
28. Meerow, S. and J. P. Newell (2019). "Urban resilience for whom, what, when, where, and why?" *Urban Geography* **40**(3): 309–329.
29. Ibid.
30. Bohland et al. (2019).
31. See for example resources from The Resilience Alliance https://www.resalliance.org/resilience
32. Rickards, L, Mulligan, M Steele, W (2019) The resonance and possibilities of community resilience, in Bohland, J, Davoudi, S, Lawrence, J (Eds) (2019) *The Resilience Machine*, New York, Routledge.
33. Bohland et al. (2019).
34. More, T (2003) *Utopia*, London, Penguin Classics.

35. Eagleton, T (2015) *Utopia's past and present: Why Thomas More remains astonishingly radical*, in The Guardian, 16th October, accessed on https://www.theguardian.com/books/2015/oct/16/utopias-past-present-thomas-more-terry-eagleton
36. Utopia is an ABC Australian comedy television series—see https://iview.abc.net.au/show/utopia
37. Gatenby, M (no date) Universities of Utopia, *Times Higher Education*, accessed on, https://www.timeshighereducation.com/features/the-university-of-utopia
38. Lewis, N and Shore, C (2018) From unbundling to market making: reimagining, reassembling and reinventing the public university, *Globalization, Societies and Education*, 17(1), p. 11–27.
39. American Academy of Arts and Science (2016) *Public Research Universities: Understanding the Financial Model*, available online at https://www.amacad.org/LincolnProject, p. 5.
40. Farbenblum, B and Berg, L (2020) "Garbage" and "Cash Cows": Temporary migrants describe anguish of exclusion and racism during COVID-19, in *The Conversation*, 17th September.
41. Marginson, S (2016) Higher education and the common good, Melbourne, Melbourne University Publishing.
42. Barnett, R (2018).
43. Waheed, H (2018) The Common Good, *The Stanford Encyclopedia of Philosophy* (Spring Edition).
44. Connell, R (2019) *The Good University: What universities actually do and why it's time for radical change*, Melbourne, Monash University Publishing.
45. Ibid.
46. Amin, A (2006) The Good City, *Urban Studies*, 43(5–6), p. 1009–1023.
47. Anderson, R (2010) The 'idea of a university' today, *History and Policy*, 1st March, accessed on http://www.historyandpolicy.org/policy-papers/papers/the-idea-of-a-university-today
48. Connell, R (2019) *The Good University: What universities actually do and why it's time for radical change*, Melbourne, Monash University Publishing.
49. Szadkowski, Krystian. 2019. The common in higher education: a conceptual approach. *Higher Education* 78:241–255.
50. Ibid.

51. Mourad, R (2020), Scholars as global change agents: Toward the idea of interdisciplinary critical spaces in higher education, British Journal of Educational studies, 68(4), p. 1.
52. Arendt, H (1958) *The Human Condition*, Chicago, University of Chicago Press.
53. Oppenheimer cited in Arendt, H (1958) *The Human Condition*, Chicago Univ Press, Chicago, p. 2.
54. Sennet, R (2008) *The Craftsman*, Yale University Press, London.
55. Sennet, R (2018), *Ethics for the city*, Penguin Books, Milton Keynes, p. 1.
56. Amin, A (2006) The Good City, in *Urban Studies, 43(5–6),* p. 1009–1023.
57. Amin, A (2006).
58. Patel, R. and J. W. Moore (2017). *A history of the world in seven cheap things: A guide to capitalism, nature, and the future of the planet*, Univ of California Press.
59. Plumwood, V. (2008). Shadow places and the politics of dwelling. *Australian Humanities Review* 44, p. 139–150.
60. Goddard, J. (2018). The civic university and the city. *Geographies of the University*, Springer, Cham: 355–373. p. 363.
61. Rickards et al. (2019).
62. Erdem, E. 2020. "Free Universities as Commons". J.K. Gibson-Graham and Kelly Dombroski (Eds.). *The Handbook of Diverse Economies*. Edward Elgar Publishing, 316–322.
63. Westendorf, J., A. Mondon and G. Hofstaedter '*How to start a free university: A guide by the Melbourne Free University*', accessed at http://freeuniversitybrighton.org/wp-content/uploads/2014/05/How_to_Start_a_Free_University.pdf, p. 4–5.
64. Marx, Karl (1993 [1953]), *Grundrisse*, trans. M. Nicolaus, London: Penguin Books.
65. See Hess, C. and E. Ostrom (eds) (2007), *Understanding Knowledge as a Commons*, Cambridge, MA and London: MIT Press; Federici, S. (2009), 'Education and the enclosure of knowledge in the global university', *ACME*, 8 (3), 454–61.
66. Cameron, J. and K. Gibson (2001), *Shifting Focus: Alternative Pathways for Communities and Economies*, Traralgon and Melbourne: Latrobe City and Monash University.
67. Gibson-Graham, J.K., J. Cameron, and S. Healy (2016), 'Commoning as a postcapitalist politics', in A. Amin and P. Howell (eds), *Releasing the*

Commons: Rethinking the Futures of the Commons, London and New York: Routledge, pp. 192–212; Neary, M. and J. Winn (2017), '*Beyond public and private: A framework for co-operative higher education*', Open Library of Humanities, 3 (2), 1–36G.
68. See Gibson-Graham, J.K. (2006), *A Postcapitalist Politics,* Minneapolis, MN: University of Minnesota Press.
69. Erdem, E. 2020. "Free Universities as Commons". J.K. Gibson-Graham and Kelly Dombroski (Eds.). *The Handbook of Diverse Economies*. Edward Elgar Publishing, 316–322.
70. Esteva, G. (2006), 'Universidad de la Tierra (Unitierra), the freedom to learn', Pimparé and C. Salzano (eds), *Emerging and Re-Emerging Learning Communities: Old Wisdoms and New Initiatives from around the World,* Paris: UNESCO, pp. 12–16.
71. Collins, M. and H. Woodhouse (2015), 'The people's free university: Alternative to the corporate campus and model for emancipatory learning', *Journal of Educational Thought,* 48 (3), 117–44.
72. Erdem, E. and K. Akın (2019), 'Emergent repertoires of resistance and commoning in higher education: The Solidarity Academies Movement in Turkey', *South Atlantic Quarterly,* 118 (1), 154–64.
73. See Erdem, E. 2020. "Free Universities as Commons". J.K. Gibson-Graham and Kelly Dombroski (Eds.). *The Handbook of Diverse Economies*. Edward Elgar Publishing, 316–322.
74. De Angelis, M. (2017), *Omnia Sunt Communia*, London: Zed Books.
75. Connell, R (2019) *The Good University: What universities actually do and why it's time for radical change,* Melbourne, Monash University Press.
76. Treanor, B (2006) *Slow University: a Manifesto,* accessed on http://faculty.lmu.edu/briantreanor/slow-university-a-manifesto/
77. Steele, W (2012) Do we need a "slow housing" movement? *Housing Theory and Society,* 29(2), p. 172–189.
78. Brown, E and McGowan, T (2018) Buen vivir—Reimagining education and shifting paradigms, *Compare* 48(2), p. 317–323.
79. Ibid., p. 317.
80. Ibid., p. 321–22.
81. *Campaign for a Global Curriculum of Social Solidarity Economy*, accessed online at https://curriculumglobaleconomiasolidaria.com/english/letter-of-principles/
82. *Ibid.,* n. p.

83. Kamola, I (2016) Situating the "global university" in South Africa, in Chou, M, Kamola, I and Pietsch, T (eds.) *The Transnational politics of higher education: Contesting the global/transforming the local*, Routledge, New York, p. 42–63.
84. Kamola (2016).
85. Brennan, J. and A. Cochrane (2019). "Universities: in, of, and beyond their cities." *Oxford Review of Education* **45**(2): 188–203. p. 188.
86. Marginson, S. (2004). "University Futures." *Policy Futures in Education* **2**(2): 159–174. p. 167.
87. Ibid., p. 168.
88. Little, B., Abbas, A., Singh, M., (2016) *Changing Practices, Changing Values?: A Bernsteinian Analysis of Knowledge Production and Knowledge Exchange in Two UK Universities, RE-BECOMING UNIVERSITIES?* Springer, pp. 201–222. p. 198.

4

Ethical Innovation

Impact and Innovation

What is the role and impact of innovation within universities today? And what has it been over time? The SDG agenda and the serious issues it points to is one reason that universities are being pushed to demonstrate their 'impact' in the world.[1] As we have seen in Chaps. 2 and 3, there are many reasons universities are being forced 'into the world', including a hunger for funding and a thirst for global status. Innovation is a go-to tool for universities in these latter quests, particularly as the very meaning of innovation incorporates real-world impact. The SDG agenda also requires innovation, but of a far more transformational sort than that which characterises standard innovation efforts. In this chapter we outline what this more innovative form of innovation might look like. In particular, we argue that it has to be consciously normative and ethical. This requires a sophisticated knowledge and understanding of the concepts, histories and social character of innovation and technology. It also requires going beyond the idea of Responsible Research and Innovation (RRI) to address far deeper questions about the worldviews at work and the worlds being created.

Two intersecting narratives tend to circulate about universities and their innovative impact. One is that universities are by definition a positive force in society and their impact underpins many advances in human civilisation as a result of their research and/or education mandate. From this perspective, universities simply need to exist to have a positive impact. The other is that, given this positive role, as well as a concurrent proliferation in the problems the world is facing and the need for universities to distinguish themselves and justify public support, universities need to increase and continually (im)prove their impact. We come then to the push for universities to be more intentionallyAnd to do that they need to become more innovative, both in the sense of generating more demonstrable innovations (real-world changes) for others and *and* in doing things working more innovatively.

These broad assessments of the situation are an adequate starting point, but they overlook one of the biggest challenges that sustainable development poses. The SDG agenda is not merely a call for universities to increase their impact, such as by expanding access or doing more public engagement work. These things are important, but the critical factor is *what impact, what innovation?* The Anthropocene demands new more sustainable and just approaches to impact. More than layering this on top of what universities already do, this requires critically reflecting on what they already do and why. Despite the positive spin that 'impact' is given in university discussions, the impact that universities have today and have had over history is far from only positive.[2] Indeed, it is intimately entwined with the problems of colonial and industrial development that have generated the need for sustainable development.

In whatever way universities are understood, they—like all other organisations and groups—are under pressure to innovate because innovation is presented as the universal solution to all problems. To the extent that innovation is understood as a verb—doing something in a different and useful way—this can make sense. Today innovation as a concept is frequently co-opted by a capitalist framing that defines it in terms of dollar value. However, beyond this narrow reading, innovation resonates with the idea of academic freedom. This is in contrast to innovation's earlier meaning as the dangerously radical questioning and challenging of the Church and established political order, which in turn was a contortion of its ancient positive meaning of 'spiritual renewal'.[3]

As universities evolved from institutions designed to study and teach religious texts to ones that started to do research—with academics consequently beginning to generate not just communicate knowledge—'innovation' evolved from heresy to a sign of originality and value. Now, as historian Benoit Godin explains, 'Innovation has become a panacea to all social and economic problems'. Perhaps reflecting the enduring psychic appeal of its original meaning of spiritual renewal, an unwavering faith in innovation has become a symbol of Western modernity. Against this universalism of innovation, Godin calls for a more critical stance. Not only does he enjoin us to go look 'critically at our theories and policies' to consider 'Is innovation the solution?', but to also ask: 'What problems does society face and what is the spectrum of solutions available or imaginable?'[4]

Strengthening the need to think critically about innovation is the way that it is now commonly taken to refer to a *thing* rather than a process. In particular, the meaning of innovation has converged with the idea of technology, such that most innovations are presumed to be technological. Since the 1930s, technology has similarly become more and more narrowly defined as the products of *techne* rather than the more expansive original Greek meaning of the study of *techne* (skilled craft or mechanical arts).[5] This conceptual reduction of innovation and technology to physical things has divorced them from their particular contexts and facilitated their commodification and evaluation in narrow economic terms.

In the absence of equal attention to studying processes, skills and craft, or offering critique or seeking spiritual renewal, this reduction, abstraction and commodification of technologies has shaped the material and social impacts that they have had in the world. In broad brush terms, these impacts are manifest in the crisis of the Anthropocene—a period notably narrated (rather breathlessly) as largely a history of sequential technological ages, with a particular focus on James Watts' steam engine as the catalyst of the industrial revolution, and the atomic bomb as the marker of the Great Acceleration.[6]

Despite the highly mixed impacts that this kind of innovation has generated, the general response to the problems of the emergent

Anthropocene has been to call for more innovation—which is to say, more of the same. Contemporary issues are inevitably positioned as a matter of too little innovation, particularly technological solutions. This 'innovation bias' pushes in a one-dimensional mode for more and more innovation (and thus impact), advancing an empty productivity agenda without any consideration of qualitative factors, context, or specific goals. Innovation bias characterises what we call a weak innovation culture because while the focus on innovation is strong, the approach to innovation and impact is not itself innovative. Rather, it is trapped within first-order reflexivity, simply perpetuating the narrative 'that innovation in the mechanic arts is a—perhaps *the*—driving force of human history'[7] rather than asking *why?* and *to what end?*

As the *unfolding of potential*, development needs to be reclaimed and remade in more qualitatively discerning—specifically sustainable, resilient and regenerative—ways. So too do innovation and impact, given their causal relationship to development. In recognition of this, there is a growing array of efforts to try to guide innovation more firmly and effectively. These critical stances on innovation and technology reflect a dawning realisation of the transformative power of our accumulating collective efforts, though not towards the techno-utopian ends vaguely envisaged since the industrial revolution.

The urgency and complexity of sustainable development means universities need to be more energetic *and* careful in generating change. In this chapter we begin looking in more detail at the question of how universities might work in a way resonant with the progressive ambitions of the SDGs. The aim is not to provide examples of how universities can engage with the SDG agenda specifically, but to explore how universities and those within them can *create the enabling conditions* needed to orient towards the SDGs. In particular, we discuss the idea of ethical innovation as a strategic route to SDG engagement. First, we explore the idea of innovation and point to the need to reclaim the concept along with that of development, calling out contemporary versions of each as implicitly capitalist and introducing the sort of *regenerative* innovation and development that the world, including universities, badly needs.

The Power, Myths and Ethics of Technology

In *The Ethics of Invention: Technology and the human future*, Sheila Jasanoff documents how innovation today is deeply challenged by questions of risk and inequality. The growing magnitude, prominence and awareness of risks and existing negative impacts, including those that have exacerbated social inequalities, are stimulating something of an ethical turn, perpetuating the emergence of what Ulrich Beck called 'Risk Society' and the reflexive modernisation that helped birth the idea of *sustainable* development (discussed below).[8] Yet, hampering efforts to address innovation's ethical challenges to date are what Jasanoff presents as three flawed beliefs which collectively give the impression 'that technologies are fundamentally unmanageable, and therefore beyond ethical analysis and political supervision'.[9]

The first of these myths is technological determinism: the idea that technology drives human progress or 'immanent development' in an inexorable fashion, beyond human control, much like a force of nature. All that human society can do, in this story, is adapt to technology as one among many exogenous forces such as 'natural disasters'. However, just as natural disasters are not natural but emerge out of the intersection of intimate relations between environmental phenomenon and society, nor are technologies (or the devastating socio-technical disasters they can unleash) a natural force. While this seems completely obvious in the sense that technologies are taken as a quintessential sign of human's distinctiveness from nature, technologies are consistently naturalised as part of our environment in an experiential and statistical sense, allowing them to fade into the background thanks to their ubiquity.

The larger and more pervasive the technology is—for instance, becoming a form of infrastructure as electricity, computing and internet technology has—the more complete this normalisation and naturalisation is, creating the sense that we humans are outside of it and have no control over it. Yet, the environmental and social consequences of treating technology in this way are increasingly unavoidable as the planet lurches into the Anthropocene on the back of the seemingly out-of-control but actually deliberately driven Great Acceleration. Among these consequences are alterations to the 'natural environment' of such a magnitude that now even natural disasters associated with phenomena such as fires or storms represent the intersection of already denaturalised, destabilised environments and society. In turn, technology's imprint on the planet is now generating challenges of such a scale that the sort of innovation we need has to go to the heart of development.

The second myth Jasanoff proposes as a barrier to effective governance of technology is a faith in 'technocracy'—elite specialists charged with dispassionately assessing and managing the risks posed by technologies. The rise of technocracy is another chapter in the rise of universities. In nineteenth-century Europe, Henri de Saint Simon advocated for a 'scientific approach to the management of society and a correspondingly authoritative position for trained experts', helping stimulate the development of the social sciences, notably economics. In twentieth-century US, there were calls to accompany technology-led progress with 'experts as advisors at every level of government'. By the start of the Great Acceleration this 'new dynamic' was so embedded that such advisors had effectively become a 'fifth branch' of government, alongside the traditional legislative, executive, judicial and regulatory branches. As Jasanoff notes, 'Scientists, nurtured by abundant public funding during the war and often relishing their role in affairs of state, lobbied hard to insert more and better science into public decisions'.[10]

Today, there is arguably more need than ever to insert robust evidence into public and private sector decision making. As discussed further below, the modern ideal of evidence-based policy is one of the drivers of the contemporary 'impact agenda' in universities. The issue, though, is that what counts as evidence, who counts as an expert, how evidence is imagined to relate to decision making, and what happens when it proves fallible, are all poorly addressed within a technocratic frame. Not only are traditional risk assessments being challenged by the destabilisation of the climate and planet more broadly—making some outcomes indeterminate (i.e. not resolvable by further research)—but such quantitative assessments are being challenged by growing awareness of the value judgements involved, leading to often fierce social contestation and declarations that science is now 'post-normal'.[11]

One result is the need to manage what Andy Stirling calls multiple 'states of incertitude'.[12] Only one such state is risk proper—that is, a situation in which both possible outcomes and the probabilities of their occurrence are known, leaving calculation the only necessary task. Other states of incertitude—uncertainty (weak knowledge of probabilities), ambiguity (weak knowledge of possible outcomes) and ignorance (weak knowledge of both)—require far more participatory, precautionary and reflexive governance (as discussed below) than the technocratic approach has traditionally provided.[13]

We come then to the third myth that Jasanoff argues is hampering good governance of technological risks, which is the myth of unintended consequences. As discussed below, this narrative of accidental, unknown and even unknowable outcomes arising from technology in aggregate is often used in representations of the Anthropocene to cast the continued and continuing fundamental degradation of the planet as unintended. For as Jasanoff puts it, 'the more dramatic' a technological failure, 'the less likely we are to accept that it was imagined, let alone intended' by those who designed it. Clearly aggregating up to the global scale sharply undermines the capacity to identify intent, responsibility or design, but such a scale also demands more than ever that we do not hide behind claims of unforeseeable outcomes. The conundrum we have to face is that growing complexity simultaneously increases the need for rapid improvements in human systems, the risk such efforts have undesirable impacts and the difficulty of identifying any specific responsibility.

Difficult as it is, this is the conundrum that the Anthropocene and SDG agenda requires us to tackle. Regardless of the practicalities, they challenge the convenient excuses that it is pointless to try and that 'it is neither possible nor needful to think ahead about the kinds of things that eventually go wrong'. In doing so, they challenge the presumption in the university sector that it is acceptable for key processes of informing, imagining and designing technologies and technological systems to remain scattered among fragmented actors and hidden from public view,[14] often behind university gates and their 'commercial in confidence' agreements with corporations. Instead, the Anthropocene and SDG agenda point to the profound need to face the need for change, identify actual and potential consequences *and* appreciate our limited control over them—underlining the need for radical innovations in approach. Unintended consequences are clearly a reality—but that is why we need to take them more seriously, not less.

Technological determinism, technocracy and strategic misuse of the unintended consequence defence are now widely challenged. Indeed, critiques of dominant assumptions about technology and risk helped stimulate the 'reflexive turn' in development—from which sustainable development and an associated 'risk society' emerged in the 1980s. More than being reflective, reflexivity questions assumptions and underlying structures and goals, including those around 'development'. Indeed, it is because of greater awareness of the costs, risks and dysfunctional character of some existing ways of doing things that the development ideal was rebooted along more careful, sustainable lines.

University research and education have an ambiguous position here. On the one hand, some have contributed to such reflexivity and helped document the many impacts that conventional development processes have had and continue to have. On the other hand, many university actors have bolstered the resilience of conventional economic development, in keeping with the push to make universities more like businesses. At the same time, ambivalence about expertise and the technocracy is one reason academics have been partially displaced by other stakeholders and voices in the ongoing transition from 'normal' (academic) science to 'post-normal' (contested, plural) science mentioned above, which aims to counterbalance narrow academic views with those of context-specific publics whose idea of risk and goals often differs starkly from the technocracy.

Whether the emergence of sustainable development in the 1980s and 'sustainable development 2.0' in the form of the SDG agenda in 2010s represents *sufficient* reflexivity is a matter of debate. Many argue that approaches to sustainable development are stuck in the mode of what Donald Schön would call 'first loop learning'—learning how to do what we already do better.[15] Like the innovation bias, or the 'master's tools' references mentioned in Chap. 1, this is about trying to fix fundamental systemic problems with the very mentalities, tools and practices that created them—retaining modernisation but making it superficially 'eco'. What is instead needed in the face of accumulating costs is more disruptive, 'second order' reflexive modernisation—Schön's 'double loop learning'—to generate a new *model* of sustainable development better suited to the systemic crises we face. From this perspective, existing norms, practices, worldviews and relationships are within the scope of what is problematised. The aim is to not only address unwanted outcomes in the world but to think deeply about how and why they have come about, including an over reliance to date on first-order reflexive modernisation.

The question, then, is whether the SDG agenda can help us do this. There is a real danger that it is too conservative. There is also a real danger that universities perpetuate this conservatism, whether out of disregard for the SDGs and the state of the world it flags, or due to engaging with it in a limited way, satisfied with merely helping solve well-paying research partners' stated problems. The issue here is not just one of insufficient reflexivity or understanding. Historians Christope Bonneuil and Jean-Baptiste Fressoz assert that experts in business and academia have long known that 'we' are pushing the Earth towards dangerous limits.[16] They document centuries of critical perspectives and marginalised voices that have tried to warn leaders and the public about the environmental costs of colonial exploitation and industrial production. In such a context, they suggest, the argument that the Anthropocene condition is unintended and only recently realised is disingenuous. Warning messages have been systematically diminished and ignored. Moreover, they continue to be so thanks to knowledge hierarchies of the sort that rank economic modelling above ecological field work, and universal theories above community experiences, not to mention corporate influences that encourage quietude or demand compliance in universities, government and the media.

Thus, the contemporary challenge is not merely to progress to second loop learning, it is to face the dark side of knowledge politics in academia and broader society, which will otherwise only continue to silence critique. As eco-feminist Val Plumwood has powerfully argued, our present-day ecological irrationality is rooted in interests and illusion as much as ignorance.[17] Nevertheless, in keeping with the university-centric bias towards technocratic risk management, most approaches to the ethics of innovation remain focused on questions of rationality and responsibility.

In a sense, this is itself rational. As Brad Allenby and Daniel Sarewitz argue, most people understand and engage with technology and innovations in a 'Level I' way; that is, at 'the level of reality of the immediate effectiveness of the technology itself as it is used by those trying to accomplish something'.[18]

In Level I, technologies are not understood as part of wider systems or longer trajectories but in terms of a highly bounded set of pragmatic questions: will they work? Who will use them? Who will own them? How do they compare to competitors? At this level, the focus is on cause-and-effect questions of the sort relatively easily answered by the right collection of experts and analyses. Ethical considerations at this 'shop-floor level' consist of normal professional concerns including adhering to domain-specific research ethics and other professional standards and codes of conduct. Although these considerations are not to be dismissed—and increasing concerns around breaches in research integrity are pushing institutions to reflect on the wider environment and pressures they are part of (including the pernicious effects of the productivity agenda in academia)[19]—overall the focus is simply on doing a good job, not questioning the job.

Things get more complex when we recognise that technologies inhabit 'another level of reality' (in fact, two). According to Allenby and Sarewitz, the Level II reality of technology is where it becomes apparent that technologies are part of 'networked and social phenomenon'. This is about reflecting on how technologies fit into broader systems, supply chains and social relations, and can feed back on those systems, helping reshape and often constrain them. As Langdon Winner asserts in *Autonomous Technologies*, and as indicated above, technologies not only help us adapt to our environment, they become our environment. As such, they stimulate processes of 'reverse adaptation—the adjustment of human ends to match the character of [technological] means'.[20] Awareness of this wider context and its frequent dysfunctionality raises questions about the systems we are contributing to, including the risk of potentially strengthening positive feedback loops between, for example, a car-centric transport sector, urban planning and fossil fuel industry, or between commodification, consumption and climate change.

Ethics here consequently needs to broaden greatly, though often it fails to do so, which means that discussion remains stuck at the 'Level I functionality of the technologies themselves', performing a 'sleight of hand' that can 'distract us from what [is] going on at a higher level'.[21] What Level II ethics demands is honest engagement with the 'social ethics' that the context of a technology brings to light, including the social changes inadvertently but still visibly generated by certain 'clusters' of intervention such as those driving, for instance, a move towards financialisation or new resource frontiers. Rather than asking whether something works, the focus in Level II is on how something intersects with the rest of society and what it therefore means for different groups, including those with little voice to shape intended impacts or call attention to unwanted ones.

The stakes are apparent when we consider the 'Fourth Industrial Revolution' (4IR): the emerging wave of industrial development characterised by automation and artificial intelligence. Many commentators assume that, being a form of development—indeed, an apparent *revolution*, one which seems to offer to further emancipate (some) humans from drudgery—the 4IR will foster sustainable development and facilitate the achievement of the SDGs, especially if it helps align the SDGs with the private sector's profit motive. For example, an Accenture Consulting report argues that 'The Fourth Industrial Revolution presents a unique opportunity to inspire innovation in advanced manufacturing while increasing companies' competitiveness and contributing to the fulfilment of the UN's 2030 Agenda for Sustainable Development'.[22] They present the main social threat of 4IR as uneven social access to its benefits. Not only does this ignore the fact that capitalism is fuelled by such unevenness, it ignores the fact that 4IR is not all positive, or even benign.

The highly mixed outcomes for employment and livelihoods, resource use, social surveillance and warfare, to name a few domains, are already becoming clear. A blog on the 'UNDP in Asia and the Pacific' warns that 'Progress on at least nine Sustainable Development Goals could be directly affected by the Fourth Industrial Revolution, and millions in Asia's emerging middle class could slide back into poverty'.[23] The blog is a commentary on the United Nations Development Programme 2018 report *Development 4.0: Opportunities and Challenges for Accelerating Progress towards the Sustainable Development Goals in Asia and the Pacific.*

While highlighting threats, it also reproduces a narrative of technological determinism, calling on countries to adapt to what is cast as an exogenous change:

> The Asia-Pacific region is in the throes of a digital transformation, accelerated by technological change, including intelligent robots, autonomous drones, sensors, and 3-D printing. Countries must adapt or be left behind if they are to achieve the Sustainable Development Goals.[24]

In contrast to this narrative of technological determinism, it is increasingly inescapable that 'technology is neither self-propelling nor value-free' and that there is a need to question the current arrangements in which 'the power to set the rules of the games for governing technology rests with capital and industry'.[25] Even the mainstream World Economic Forum calls on boards of directors to 'ensure that their firms are creating *long-term* economic value and not just short-term financial returns' and, more significantly, strive to ensure that the 4IR creates 'more equitable economies' via 'much more human-centred' economic, labour and education policy.[26] They even suggest that society may need to rethink what constitutes economic value.[27] This shift towards meta-ethics underlines the potential for the 4IR-SDG relation to help stimulate discussion about the big questions underpinning both: questions about what is meant by development and who it is for. If universities do not engage with these broader challenges, caught instead in an anxious race to fill Level I reality with more innovations, they will badly underestimate how the world is changing.

A more fundamental challenge again stems from what Allenby and Sarewitz call the 'Level III reality' of technology. More than technologies-in-use, or technologies in large infrastructural and social systems, this is about technologies in the Earth System. It is about their emergent collective effects on and through the rest of the planet, and the resultant, almost chaotic dynamism between its interacting parts. The ethical stance needed here, Allenby and Sarewitz assert, is a 'macro-ethics': a process orientation that strives to adhere to basic principles such as the Precautionary Principle (discussed below). It means accepting that unpredictable outcomes require 'muddling through' and questioning even ethical precepts and worldviews. The core ethic required here, the authors assert, is a simple commitment to continuously engaging with the challenges—to 'staying with the trouble' as Donna Haraway might put it, and which we discussed in the previous Chap. 3.

Allenby and Sarewitz especially emphasise the need to continuously engage with the institutions and systems we have collectively established to manage these risks on our behalf:

> Because macro-ethics requires ongoing dialog with systems that are changing unpredictably, and in many dimensions (technological, social, natural, ethical, and economic, among others), individuals should support constant institutional engagement with such systems.[28]

Sheila Jasanoff similarly emphasises that technologies shape not just our physical or economic worlds but our worldviews and imaginaries, our social expectations, norms, practices, laws, hopes, fears, institutions and intentions. And she similarly concludes that to manage 'our grand bargain with technology' what is needed above all else is 'deeper ethical and political engagement'.[29] This institutional engagement very much includes universities, both in terms of individuals engaging with them, and through them to other wider institutions. For universities to not engage is to give up the chance to speak to neglected risks, highlight marginalised voices and make visible unspoken assumptions.

In direct and indirect ways, the SDG agenda and the challenges it draws attention to is a key feature of our shifting environment. Yet it can be easily ignored if we bunker down in Level 1 reality, as some university staff—among many others—are well-practiced at doing. But to do so would be deeply self-defeating. So too would dismissing it because it is imperfect. The question is not whether it is perfect enough to be worthy of our engagement, but how we can make it better through creative implementation. As Professor of Global Economic Governance Ngaire Woods warns, 'without bold innovation, the new development agenda will be far from sustainable'.[30]

We now introduce two frameworks for thinking further about the sort of innovation universities need to help foster for 'the rest of the world' and in their own institutions and sector. The first of these frameworks is an extant one in Europe that points to positive moves to more genuinely engage with ethics, and the second one is an original one—an innovation if you like—that we have designed to push the conversation further in keeping with the unfolding Anthropocene.

Responsible Research and Innovation

Among the innovations needed to achieve the SDG agenda are innovations in our policies and the governance of innovation itself. This is precisely what the co-emergence of the Responsible Research and Innovation (RRI) framework in the European Union represents.[31] First discussed in 2011 and now incorporated into European research policy alongside an explicit focus on contributing to the SDGs, RRI is a particular formalisation of older but partial initiatives to assess the legal, social and ethical implications of technologies such as genomics.[32] As Richard Owen and colleagues have discussed, it is the most comprehensive and ambitious effort to date to shift the focus from 'from science in society, to science for society, with society'. *Science for society* focuses attention on the purpose of innovation and tries to target it at the 'right impacts'.[33]

The SDG agenda is a useful, albeit incomplete, guide to these 'right innovation impacts', noting that such impacts often consist of the negation or management of the wrong (sometimes unintended) human impacts on the world, for instance climate change. The ethos of the SDG agenda—notably its focus on reducing inequalities and good governance—makes it relevant to RRI's focus on *science with society*. This is about:

> the need for research and innovation to be responsive to society in terms of setting its direction, and in modulating its trajectory in the face of the uncertain ways in which innovation invariably unfolds as part of its naturalisation in the world.[34]

Explicit about the undemocratic character and failures of past innovation policy and the need to guide innovation to come, RRI frames 'innovation as a future-oriented, uncertain, complex and collective endeavour'. It challenges 'scientists, innovators, business partners, research funders and policy-makers to reflect on their own roles and responsibilities' and acknowledge that past and present 'irresponsibility in innovation is a manifestation of the ecosystem of innovation and requires a collective, institutionalised response, if this is indeed possible'.[35] By calling out irresponsibility, RRI is a potentially radical and subversive vehicle for addressing some of the most entrenched structural problems of the Anthropocene, including the spatial distribution of environmental goods and harms along wealth lines of the sort that underpin the continued justification for further socioeconomic development. That is, it pushes researchers and others to engage with Level II and Level III realities.

To try to foster the sort of collective response possible, a somewhat watered-down version of RRI is now embedded in the European Commission's flagship Horizon 2020 research strategy and is the subject of many outreach efforts designed to embed it into universities and other settings. Its overall framework envisages the 'research community' and 'education community' (e.g. universities) working with policy-makers, businesses and industry and civil society organisations to collaboratively govern innovations using mechanisms at various scales (e.g. from individual researchers' practices, to open data access arrangements), guided by four principles:

- Diverse and inclusive voices in decision-making
- Open and transparent decision-making
- Anticipatory and reflective assessments
- Responsive and adaptive measures.

Despite not directly addressing previous irresponsibilities in innovation and their still-unfolding effects, this institutionalisation of RRI is a valuable step forward when it comes to harnessing innovation and universities to the SDG agenda. In particular the focus on science for society and generating the 'right impacts' aligns strongly with the use of the SDGs as a research agenda. At the same time, the significance of RRI lies in the way in which it applies SDG-like principles to the process of research and innovation. That is, it illustrates the sort of outcome that can emerge if the SDG focus on good governance, social and environmental risks and needs, gender equality and education are directed 'inward' at the innovation production process and universities themselves.

RRI is thus emblematic of the sort of implications the SDG agenda could have for universities and the broader research world. John Goddard argues that an RRI approach encourages a civic university model because both try to make visible the real people and places affected by interventions.[36] Despite much talk of RRI, recent analysis of its application across 12 nations indicates its uptake has been patchy, partly because of tensions between the research productivity and excellence agendas and a dawning realisation of what it actually takes to make 'room for adherence to RRI'.[37]

While there is a shared interest in positive real-world impacts between RRI and SDG work, the two are not equivalent. As a universal tool, RRI theoretically applies to any research topic. This agnosticism means almost anything (e.g. innovation in coal combustion, chemical warfare and cigarettes) could be endorsed by it, if scoped in a certain way. It does not call research goals and areas into question, only the way in which work on a given topic is conducted.

Although RRI is being used to drive and guide SDG-oriented innovation—for example, generating new business models and business-academia partnerships[38]—this alignment is not intrinsic to it. The model of responsibility promotes some issues highlighted by the SDGs, notably gender equality and partnerships, but leaves out principles that are essential to addressing the fundamental challenges of the SDGs in the Anthropocene. These challenges not only push innovation further towards the SDGs but also expose the need for radical innovation within the SDGs and approaches to them.

RADAR: Beyond RRI to Ethical Innovation

Building on and deepening RRI, an ethical innovation approach for the Anthropocene encompasses a far wider scope of innovation issues and more expansive understanding of technology. Encapsulated by the acronym RADAR, our ethical innovation framework emphasises that rather than ethics being a single, separate element of innovation processes (as in RRI), ethics and the associated politics infuse the whole arrangement and context of innovation including their shaping of the world, development and us. In doing so, we follow Kirsten Jenkins and colleagues in trying to bring RRI into conversation with other broader discussions about the ethical dimensions of technologies and development, notably the need for justice.[39] We argue that not only distributive and procedural justice, but *cultural recognition justice*, which questions the dominant hierarchies that structure knowledges and society (that is whose knowledge counts) needs to be part of an ethical innovation model fit for the present.

Our resultant framework underlines not only the vital need for Responsibility of the sort discussed under RRI, but more reflexive responsibility, as well as a commitment to Attentiveness, Disruption, Authenticity and Regeneration (see Table 4.1). In this way, our Ethical Innovation framework helps free discussion of innovation from the technocratic mode that RRI is still arguably stuck in to ask the bigger questions that impending planetary collapse demands. While it overlaps with technology assessment processes, it applies to any stage and site of innovation. The focus is on universities and what the SDG agenda and its gaps demand of us, including the need to acknowledge and redress the agenda's weak inclusion of Indigenous people and their perspectives.

The five-part framework is meant as a list of provocations rather than a prescriptive guide. As critical geographer Tariq Jazeel notes in his book *Postcolonialism*, 'there can be no easy or proscriptive step-by-step manual for responsible … knowledge production because each situation is particular'.[40] Thus, only what Jazeel calls 'strategic tactics' can be offered.

Responsible Innovation

As RRI indicates, responsibility is the classic starting point for discussions of ethics, and it remains a core pillar in our framework. Professional ethics, including formal research ethics processes, play an important role in universities in averting harms and cultivating a moral ethos. Yet, 'organizational routines both produce and diffuse concerns about the risks and benefits of scientific research and products'.[41] Formal ethics codes and procedures can distract from the broader ethical relations that activities and people are part of.[42] By regularising processes and requiring compliance, such procedures can shut down rather than open up consideration, conversation and learning about issues at the heart of universities.

Table 4.1 The RADAR framework of ethical innovation

Characteristic	Description
Responsible	• Uphold professional ethics and standards and engage with them in a way that fosters understanding of the need and purpose of them and the counter-veiling pressures they face • Ensure innovation processes are inclusive and foster procedural justice • Adopt a precautionary and complexity-aware stance on an innovation's potential effects in the world • Ensure that benefits are shared between all involved and that relationships are accountable and reciprocal
Attentive	• Be reflexive about assumptions and the limitations of our theories • Track effects and impacts carefully and adjust course as needed • Foster cross-disciplinary exchange, epistemological humility and a learning orientation • Address knowledge politics including a bias towards quantitative knowledge • Avoid options that close down possible positive futures
Disruptive	• Carefully question the value of what is proposed and consider assumptions and alternatives • Prioritise innovations that are most likely to generate crucial positive change, noting this might not be the most spectacular option but may be, for example, ones where the greatest learning takes place • Look for opportunities to defamiliarise the taken-for-granted and rethink common sense • Appreciate that some things need protection from innovation and other disruptive processes, requiring instead careful nurturing • Address the need to disrupt and dis-embed some existing innovations
Authentic	• Ensure innovation is driven by a genuine desire for positive change and a well-informed understanding of a situation • Be alert to innovation bias and the distorting effects of institutional struggles and pressures, including the research excellence and productivity agendas • Be alert to the ambiguities, ethical complexities and opportunities of working with others, including partners and participants with different expectations, priorities and interpretations of problems • 'Abide by' the places, communities and things that you work with and create over the long term and avoid extractive relationships • Respect incommensurable differences and avoid misrepresenting or co-opting others' knowledge

(continued)

Table 4.1 (continued)

Characteristic	Description
Regenerative	• Be committed to not only avoiding harms but generating positives, to not only investigating the world but using every encounter and stage to foster good • Use the innovation process to bear witness for marginalised or silenced groups and beings to try to heal and prevent cultural misrecognition • Be committed to repairing past harms and injustices including the cultural misrecognitions that structure colonial knowledge hierarchies • Be aware of the potential and need to generate positive outcomes at all stages and levels of a process and commit to giving back to participants in an appropriate and sustained way • Engage with peers and the Academy in a way that sustains and grows the most precious elements and avoids enhancing those that are degenerative

Besides Level 1 ethics, responsible innovation has to incorporate Level II realities. This is about thinking through not only intended effects but possible negative ones. Although formal risk management is standard practice in most projects, often the focus is on risks to the institution. But the real-world orientation that the SDG agenda encourages means focusing more intently on those we engage with through academic work or who may, in a future and indirect way, be affected by it. This begins by attending to who we welcome into our work within our institution and work environment, and considering what role we invite them to play, what support we give them to play it and who is left out.

If, for example, SDG-oriented efforts in universities are conducted in a way that is blind to the agenda's intent around promoting decent work (SDG 9) and reducing inequalities (SDG 8), the result is not just hypocrisy and irony, but a perpetuation of unjust and unsustainable development, thus undermining the initiative's contribution to the global goals. The same is true of work that generates other harms through its often invisibilised processes, including unnecessary and unjustified resource consumption, greenhouse gases, plastic waste, biodiversity degradation and so on. Academic work is a situated, material practice that cannot escape the shared world highlighted by the SDGs or dodge the implications of the agenda's exposure of unsustainable production and consumption practices.

In addition to any side-effects generated by an innovation process, responsibility means mapping out the possible impacts the given effort eventually has when 'released' into the world. Although reading universities' Impact Case Studies gives the impression that research does not generate negative effects, it is foolish to think that only positive impacts are generated. Decades of work in international development underlines the serious risks that poorly considered or consulted interventions generate. Regardless of what context we are working in, there is a need to take on board the hard-won lessons from these development practitioners and academics as to what *not* to do.

Top of the list are reductionism and imperialism. Both fail to appreciate an empirical context in its real, complex particularity and instead impose a dangerously narrow and arrogant lens that ignores or misrepresents crucial elements, including local actors and their deep knowledges. While not all university work is based on an empirical context in the way international development research and innovation is, all of it has an eventual empirical element by virtue of being part of the world. Thus, as author of *Another Science is Possible,* Isabelle Stengers, puts it: 'No decision is ever innocent'.[43]

The SDG agenda is also not innocent. For one thing, it marginalises Indigenous people's voices, as many have pointed out. When engaging with the SDG agenda we thus have a responsibility to do so in a way that strives to acknowledge, respect, understand and witness Indigenous peoples and their sovereignty. In the case of research, this means engaging with Indigenous research ethics. While all Indigenous peoples have their own unique perspective, scholar of research ethics Helen Kara suggests that four principles are often present:

- *relational accountability* (understanding research as a web of relations, including with 'nonhuman' others, and that the relationships—even if initiated for the purposes of research—exceed and take priority over the formal research);

- *communality of knowledge* (recognising that everyone has valuable knowledge to share and that research knowledge, being inherently relational, belongs to everyone and everything);
- *reciprocity* (understanding the relationships involved in all elements of research, including with the broader environment, as ones of mutual obligation);
- *benefit sharing* (the principle that any benefits from research should be widely and equitably shared with participants and communities).

Attending to Indigenous research principles resonates with key elements of justice in the Western canon: the distribution of benefits and harms; inclusion and fairness in the procedures and decision making; and cultural recognition of and esteem for all groups involved.[44] It also closely aligns with an Egalitarian worldview. In other words, despite their fundamental differences, there are strong synergies between Indigenous and Western ethical perspectives and a real opportunity for the latter to learn from the former, which would in and of itself constitute a significant, positive outcome.

How the effects of an intervention affect different people, places and periods remains a profoundly difficult question. Our responsibility to be as vividly aware of potential consequences as possible can be intensely anxiety-inducing when combined with the deep uncertainties that increasingly cloud the future in the midst of Level II and Level III complexities and a heightened sense of relationality and responsibility. Andy Stirling proposes that in the face of risks, uncertainties, ambiguities and ignorance, the best we can do is adopt the principle of precaution, which means assigning the benefit of any doubt towards the desired goal, or *goals* in the case of the SDGs. That is, 'what is better for the SDGs?' becomes the rule on which decisions are made.

Of course, answering even this question is immensely difficult, and Stirling underlines that overall being precautionary is about approaching all such situations as a social learning challenge, one in which the inputs to the decision (e.g. knowledges and voices, considerations) are deliberately kept as wide as possible and so too are the possible outputs or options under consideration. That is, what an intervention could and should do is approached as a fundamentally open question more than a quest to lock onto a 'right' answer.

Attentive Innovation

We come then to the overlapping idea of Attentive Innovation, by which we mean innovation processes that are reflexive about assumptions and attend to emerging feedback/s adapting course as necessary. Allenby and Sarewitz argue that the Level III realities of innovation today—that is, the context of a dynamic, increasingly unpredictable Earth System—mean that we should eschew the idea of clear 'solutions' (which is a Level I notion) and instead use scenarios and other anticipatory and plural techniques to explore possible outcomes, then 'lower the amplitude and increase the frequency of decision making' to muddle through adaptively.[45] While this approach can seem 'inefficient', they argue it is more effective and thus more efficient in the end.

The need to be attentive demands innovation processes informed by perspectives from different academic disciplines and real-world experience. Combining disciplinary perspectives like pieces of a jigsaw puzzle provides a more comprehensive knowledge of the sort needed to design, anticipate and monitor innovation outcomes. Beyond such multi-disciplinarity, integrating disciplinary perspectives more fully into new interdisciplinary knowledge can provide deeper understanding, including crucial self-awareness among participants of the particularities and limitations of their own discipline's perspective.

This requires attention to the learning process, and an appreciation of interdisciplinary endeavour 'not as a grand project, routinely advocated for but rarely delivered, but as a series of negotiations and recursive interactions between disciplinary perspectives', as geographers Judith Petts and colleagues conclude from their study of UK researchers working on urban sustainability.[46] Petts and her colleagues argue that this requires not only the space and time for sharing ideas and knowledge, as well as mutual trust and respect among those involved, but also agreement that 'the problem' under investigation can be framed in different ways.

The idea that people 'frame' the world at all and do so differently is the central pillar of what is called an interpretivist epistemology, where epistemology refers to what is true and interpretivism refers to multiple interpretations of the world.[47] A consciously 'interpretivist stance' can greatly facilitate interdisciplinary collaboration and thus attentive innovation. It encourages us not only to reflect on who is included in a project team but to include that a social scientist or someone else who regularly works with an interpretivist epistemology is involved so that they can facilitate the group's knowledge sharing, translation and integration process.[48] Those working from more of a positivist perspective are often unfamiliar with the sort of discussion-based learning and inquiry processes involved in interdisciplinary collaboration and can be resistant to the 'epistemological humility' that it can require.[49] Associated with qualitative research, interpretivism can also valuably bring into a project others' perspectives, such as the concerns of potential 'end users' or local communities, which are crucial to enacting Responsible Innovation.

Being attentive also pushes for deeper awareness of the knowledge politics that interpretivism exposes and is entangled in. As the idea of cultural recognition justice (mentioned above) underlines, not all knowledges are interpreted as equally valid. The knowledge hierarchy of Western modernism has a pernicious effect. In particular, Indigenous knowledges are consistently dismissed as mythical or of lesser importance than conventional academic knowledge. Within academia, the irony is that qualitative research of the sort that highlights different interpretations and the politics of their relations is regularly devalued as less true or useful than more quantitative research.[50]

Quantitative research is advantaged by knowledge practices that foster success on all three of the major meta-agendas at work: supporting universal knowledge claims of the sort celebrated in the research excellence agenda, allowing a single epistemological starting point that enables team science and high levels of research productivity, and encouraging the application of knowledge to technological innovations that are readily recognised as impactful. In this way, a positive feedback loop between quantitative research and success in academia's meritocracy has been established, illustrating what systems thinker Donella Meadows calls the flow of 'success to the successful'—where a certain group's success allows them to structure the 'rules of the game' to foster their future success.[51]

Interdisciplinary work is increasingly valued as a means of fostering new knowledge at the interface of disciplines, and better fitting academic research to the messy contours of real-world problems. This has increased the attention afforded to HASS disciplines including qualitative research.

However, the form of cross-disciplinary collaboration sought is often of a 'subordination-service' mode rather than more equitable 'integrative-synthesis' mode, to use Andrew Barry's and colleagues' terms.[52] In such a mode, qualitative researchers are included not to coproduce new knowledge and innovations as much as to manage stakeholders, translate the results for funders and other audiences, and facilitate team dynamics.

This misrecognition and misuse of qualitative research is part of what needs to be attended to in ethical innovation. Acknowledging the substantive value of interpretivist research helps address this unjust knowledge hierarchy. Moreover, valuing and adopting an interpretivist stance can help others become more humble and self-aware about their own epistemology, helping trigger learning and open up innovation processes to *multiple* different perspectives in the world. Attending to other knowledges in this way facilitates an improved understanding and adaptation of the actual outcomes of an innovation process. It is partly because communities' experiences of and concerns about innovation-led development processes, as well as warnings and evidence from marginalised 'impact sciences' such as ecology have been ignored by those caught in and blind to knowledge hierarchies that the world is now in such an unsustainable state. Making innovation more attentive to these historical injustices and biases is crucial to making it more sustainable in the future.

Attentive innovation also needs to be rooted in keen awareness of the limitations of our theories and assumptions, regardless of discipline or epistemology. More than conceptual, this is about asking how our intellectual building blocks—including those that exceed our disciplinary specialisation—allow us to attend to and engage with the world around us. For example, do we have 'a theory of society con*ceived with practical intention*'? as Habermas called for, echoing Marx.[53] Or are our theoretical lenses lacking in intention, giving new meaning to the idea of unintended consequences? To generate intellectual concepts fit for the world around us, we need to 'learn to be affected' by that world as Bruno Latour has argued. Do we, for instance, approach our work with 'a posture of openness, of welcoming, of invitation, towards earth others'? Or, stuck at our desks within our carpet worlds, do we adopt 'a stance of pre-judged superiority, of deafness, of closure?' and thus reinforce the subject-object dualism between humans and the nonhuman world that has structured Western thought?[54]

Critical human geographers J.K. Gibson-Graham and Gerda Roelvink argue that 'Performing this dualism has arguably led us into planetary crisis, and "un-performing" it may turn out to be a key practice in an ethics for the Anthropocene'.[55] This idea that an ethics for the Anthropocene requires that we 'learn to be affected' and adopt a non-dualistic, relational outlook further reminds us that we need to learn from Indigenous people's knowledges and relational ethics.

Like some of the alternative approaches to development discussed in Chap. 2 and the alternative forms of university and education discussed in Chap. 3 (e.g. the Education and Social and Solidarity Economy Network), J.K. Gibson-Graham's scholarship pushes us to rethink economies and open our eyes 'to projects and possibilities of non-capitalist development here and now'. It helps attune us to 'what is already being done', providing a starting point for the more common question of 'what is to be done?'.[56] Attentive innovation is about this sort of opening up of the 'option space'—the range of ideas and potential pathways available—as Stirling above, and Allenby and Sarewitz advocate. Having alternatives, including alternative conceptual lenses, is key to avoiding the sort of 'lock in' that characterises the Great Acceleration period of the 1950s that turbo-charged the Anthropocene. As the uncertain conditions that the latter is now contributing to intensify, the search for multiplicity, niches and positive alternatives, initiatives inspired by the SDGs are helping provide possibilities, including within universities.

The continual potential for perverse outcomes means that layered across efforts to proliferate options needs to be a critical evaluative frame of the sort that careful qualitative research can help develop. For not all options are good ones or acceptable to all. Some options are what designer Tony Fry calls 'defuturing': closing down possible positive futures, often in pursuit of short-term gains.[57] Rather than adopting such options, we need instead to deliberately turn away from them so that other futures may flourish. Opening up options is not a mindless relativistic exercise, but one that requires us to identify which pathways seem to hold open the most positive future options. It is about carefully shaping the option space, not just growing it larger. We return then to the need for a qualitatively distinguished form of development, one that takes development's central idea of 'unfolding potential' and adds critical evaluation of that potential—as the SDG agenda seeks to do.

The Anthropocene has already demonstrated the dangers of an uncritical approach and the fact that some potential, once unfolded, cannot be 'refolded'. Taking the lessons of the Anthropocene to heart and adopting the precautionary stance discussed above means attending to not only what needs to be cultivated but also what needs to be ruled out, prevented and dismantled.

Disruptive Innovation

The urgency and stakes of the Anthropocene mean that innovation choices are more significant than ever. In making such choices, the evaluative criteria in use need to include the importance of being disruptive. By adopting the term disruption, we are trying to highlight first, that an innovation needs 'to work'. Given the time pressures on the world, we need to at least aim for interventions that are effective, including cost-effective, where costs are considered not only in a pragmatic Level I frame (see above), but in terms of long term social and environmental outcomes, in keeping with the Level II and III contexts of technologies. Many innovations being pursued today are ineffective because they are not costed to take into account the greenhouse gas emissions and environmental damage generated throughout their life cycle, or do not factor in climate change impacts and so are vulnerable to its increasingly obvious physical and flow-on effects. In other words, they are a waste of time and money because they are ill-fitted to the world they are meant to improve and worsen that fit by eroding the stability of the planet. Others are a waste of time because they will never be widely or properly adopted thanks to failing to take into account the social context and others' perspectives (of the sort that qualitative social science could have helped illuminate). Anthropocene history is full of innovations that have failed to be adopted widely or permanently or as intended, or that have generated immense, expensive, unjust social and environmental harms because of the insensitive and imperial way they have been 'rolled out'.

The problems generated by the sometimes violent dissemination of 'successful' innovations raises the question of what counts as successful, effective and disruptive. Important here is not only the time scale and breadth of evaluation, but the criteria of evaluation. In other words, what is the goal? Moving the goal posts is one of the most effective routes to disruptive innovation. In the terminology of systems thinking, changing goals is a powerful 'leverage point' for intervening in—and disrupting—a system. Systems thinker Donella Meadows' famous Leverage Points framework explores the effectiveness of different tactics for changing systems. She rates the changing of system goals as more effective than numerous other approaches, including altering system parameters such as indicators, or disrupting positive feedback loops so success does not only flow to the already successful. Changing goals is also more effective than increasing negative feedback loops by adding in 'checks and balances' (for example thresholds of harm that technologies should not exceed).

As David Abson and colleagues explain in their application of Meadow's ideas, changing goals is effective because it prompts changes to the design of a system, which combined with initiatives that change the *intent* of a system (discussed below), are the most effective approaches to generating real change (Fig. 4.1). It is important to note here that by 'goals', Meadows is referring to whole-of-system goals. Rather than referring, for instance, to a project goal such as maximising the profitability of Innovation X, the focus is on larger underlying goals such as economic growth or the whole idea of human development. System goals are about 'the point of the game', which (in keeping with the meeaning-in-action idea discussed in Chap. 1) Meadows notes are 'not so much deducible from what anyone *says* as from what the system *does*'.[58]

The SDGs can be thought of as a potentially disruptive innovation in that they are an attempt to change the goals of development. Whether they are effective as such an intervention is an open question. Arguably the state of the global system indicates that, despite what the SDGs *say*, the dominant goals are still driving towards unsustainable development. Nevertheless, the SDGs are important from a systems perspective in exposing the existing goals and opening them up for debate. They also arguably contribute to a range of 'lower level' leverage points: increasing awareness of the power of human systems to self-organise (i.e. the fact

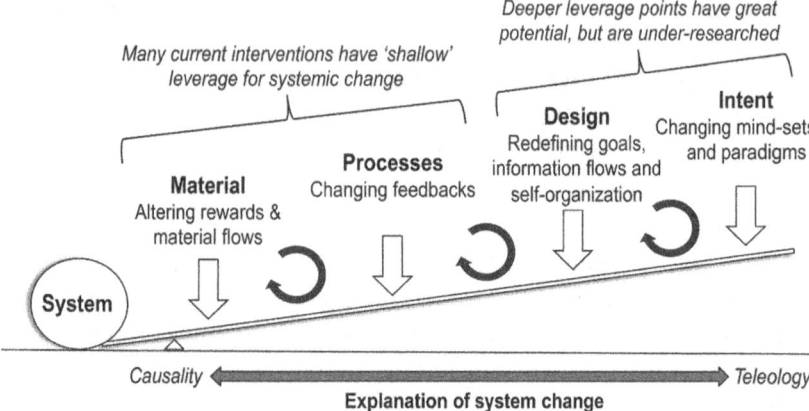

Fig. 4.1 Leverage points for systems change. (From Fischer, J., Riechers, M. (2019) A leverage points perspective on sustainability. *People and Nature* 1, p. 115–120)

that we could actually do things differently), changing the 'rules of the game' (e.g. via the TIMES Higher Education Impact Index ranking of universities) and, in turn, altering informational flows and decisions.

Furthermore, the SDGs are arguably prompting awareness of, and discussion about, the most powerful considerations or system leverage points: the paradigms and worldviews that shape how the world is understood, and the need to challenge and diversify them. While the SDG agenda does not in itself disrupt dominant worldviews and introduce alternatives, by beginning to call into question system goals such as non-sustainable capitalist development and its bias towards any and all profitable innovation, it offers a pathway towards doing so.

Authentic Innovation

The fourth characteristic of ethical innovation we propose is authenticity. This takes the idea of intent and considers it from the perspective of virtue ethics. Virtue ethics are ethical frameworks that encourage ways of acting that express virtues such as honesty and wisdom and discourage ways of acting that express vices such as greed or malice. Advocated by Aristotle, virtue ethics draws attention to the internal drivers of our actions. Arguably the SDGs are more strongly shaped by a utilitarian ethical framework focused on outcomes and a deontological framework focused on rules than by virtue ethics. While a focus on actual outcomes and rules-in-use is vital, so too is a virtue ethics approach highly relevant to thinking through the SDG agenda. This is for three reasons.

First of all, the complexity of the world means that we cannot rely on our capacity to detect and adapt to emerging outcomes, as a utilitarian ethic presumes. Even combined with useful deontological rules such as the precautionary principle leaves too much at stake. We need a virtue ethic to further increase the odds of desirbale outcomes. Second, a virtue ethics approach emphasises the importance of not just intent but its relationship with habits and actions. Rather than feeling despondent about how imperfect we are—that is, how far we are from being what Aristotle called an 'excellent person'—he encouraged us to act as if we are as virtuous as we wish we are. This 'fake it till you make it' approach is supported by social norm theory that suggests that changing statistical norms through top-down measures such as laws can drive changes in injunctive (social) norms—people's beliefs about things 'ought to be'. Although the SDGs are not enforced rules, if they communicate a shift in what is socially acceptable and encourage changes in practice, those people altering their practices may come to agree with the new way of doing things, even if they did not start out with a genuine commitment to the goals. Indeed, engaging with the SDG agenda offers a potential pathway to more authentically engaging with not only the agenda itself, but the question of ethics and their relevance to development and universities.

Second, the blurry line that exists between thought and action, and individuals and their social context, highlights the relational quality of virtues and the value of cultivating positive enabling conditions for them to be expressed. The role of this environmental influence reminds us that how we coproduce the SDG agenda in practice depends not only on 'the agenda' but the complex context we are also part of. In turn, this highlights the potential importance of institutions such as universities. It is significant that contemporary universities may *not* as yet offer a positive enabling environment for virtue ethics and their expression through the SDGs. Arguably universities are sites in which there is considerable risk that people engage with the SDGs superficially, without any genuine understanding of or commitment to them, which is why we emphasise the question of authenticity. Simply following the letter of the law or 'chasing the indicator' rather than acting in accordance with the spirit or intent of the SDGs can lead to perverse outcomes.

Unfortunately, numerous characteristics of contemporary universities—including excessive competitiveness, a commitment to growth and a preoccupation with brand and reputation—encourage such an approach. Encouragingly, though, this underlines again our point that limitations with 'the SDGs' are generated as much by how people engage with them as the agenda itself. Arguably the overall intent of the SDGs is genuine, positive, social-environmental transformation, even if the precise goals, targets, indicators and examples of implementation to date do not live up to the expectations. We certainly believe that it is possible to read the SDG agenda generously and that its actual effects in the world depend on how it is co-produced in practice. These effects include the opportunity to expose and help redress serious pre-existing issues within universities such as competitiveness and cynicism.

Regenerative Innovation

We come then to our final criterion of ethical innovation which speaks to our attempt to engage with the SDG agenda generously but critically, calling out the agenda's weaknesses while trying to compensate for them through creative, innovative, *regenerative* engagement. These weaknesses include a relative absence of regenerative approaches within the SDG agenda, though some of the work they are stimulating on the ground demonstrates their regenerative potential. By regenerative we mean invigorating, nurturing, healing. As captured in the term regenerative agriculture—an increasingly popular form of farming directed at restoring the soil and cultivating thriving social-ecological systems—regeneration is now recognised as an important ethic in the Anthropocene. Acknowledging the harm the Anthropocene has already generated, regenerative approaches build on and move beyond harm-minimisation approaches such as risk management to try to heal past harms and generate new positive outcomes.

Regenerative innovation involves continuous 'reparative practices of knowing',[59] ones that understand the inseparability of knowing and doing, individuals and context, research and impact. Feminist scholarship has long emphasised the many ways in which research is 'generative', including but not limited to any specified activities designed to disseminate or implement certain end products in the world. Research generates effects from the moment it is conceived. Beginning with 'bringing its subjects into being', research—and all knowledge generation processes including teaching and learning—help bring the world into being.[60]

By inventing or legitimating some realities and not others, and being shaped in turn by those, knowledge production helps co-produce the world. With the world now in an increasingly de-generate state, there is an overdue need to critically evaluate this power and responsibility. In particular there is a need to examine how knowledge generation within universities has helped generate and is continuing to generate the current world, from micro to macro scales, and thereby explore how it could re-generate more habitable and humane ones.

The ways in which universities have reinforced unhelpful approaches and often failed to support positive alternatives is illustrated by the case of economics, which is both a discipline within universities but also a discourse that universities operate within and reinforce through their own practices. More than just discursive, classical economics is an evaluative lens that strongly shapes our physical and social world, generating some socio-technical arrangements and not others, helping manage some negative impacts and not others. As Gerda Roelvink puts it 'the power of economics lies precisely in its ability to produce material effects'.[61]

Universities do not just continue to generate and teach classical economics knowledge, they foreground it in their own decision-making, including real estate and investment decisions. In doing so, they implicitly background not only non-economic considerations such as staff and student morale or campus biodiversity, but—with some notable exceptions and alternative models (see Chap. 3)—they help marginalise alternative economic paradigms such as degrowth and the other counter-hegemonic ideas of ecological economics.

Given this, it is of little wonder that something as debated, negotiated and necessarily compromised as the UN SDG agenda contains a reference to the central goal of classical economics—that is, economic growth. But rather than reject the agenda for this reference as some critics do, we argue that we need to see the consistencies between it and the university system that many critics are housed within, acknowledge the need for change across the board, and seize the opportunity the SDG agenda provides to progress such change. By being explicit about economic growth—notably through *SDG 8 Decent Work and Economic Growth*—the SDG agenda helps reveal what is usually privileged but implicit. As such, it helps stimulate much-needed dialogue about alternatives and provides a platform for what this would mean for other goals as diverse as dealing with ocean pollution, maternal health or disaster resilience. Turning to alternative economic paradigms and goals and exploring how they can be integrated with the many other positives that the SDG agenda represents becomes a central task.

There is an important opportunity here for another far-reaching regenerative 'innovation'—to turn to Indigenous groups and learn from and support their worldviews and approaches, including their alternative understanding of economics. Not only would this help to heal some of the profound damage and injustices that capitalist economic development has wrought upon them, but it would help all of us find productive alternatives to such development.

For example, numerous commentators have noted the regenerative potential of turning from economic growth to the Indigenous philosophy of *Buen vivir* which translates roughly into 'living well'. Standing in contrast to what is cast as *malvivir* (a bad way of living) and *maldesarrollo* (bad development), *Buen vivir* is a pluralistic and multifaceted approach that emphasises material sufficiency, radical democracy and genuine equality. It celebrates Indigenous sovereignty and culture and the physical and ethical inseparability of humans and the nonhuman world. The latter is understood as Pachamama, or Mother Nature,[62] pointing to potential synergies with certain Western philosophies and paradigms such as Aldo Leopold's land ethic, James Lovelock's Gaia theory or the symbiogenesis (cooperative evolution) that Donna Haraway has influentially explored.

Buen vivir's relational ethic has far-reaching implications for how development is understood and universities approach education.[63] While some argue that it is inherently opposed to the very notion of socioeconomic development, others see it as complementary, as a way of altering what systems theorists such as Donella Meadows would call the very intent of the system. Although in some ways alien to the SDG agenda, a number of authors have noted the resonance between *Buen vivir* and the

SDGs.[64] There is potential to use these synergies to open up spaces in the SDG agenda for *Buen vivir* thinking and practices, cultivating niches within the agenda for more regenerative approaches. In doing so, the SDG agenda's existing pluralism would be greatly expanded, helping surface and weave together worldviews in creative ways, and thus reaching what Meadows suggests is *the* most powerful leverage point in a system.

The ethical innovation needed in implementing the SDGs has to begin with acknowledging its contradictions, gaps and possibilities because these limitations are manifestations of the plural process and worldviews that gave birth to it. Engaging with it is to engage with a wide and crucial conversation about the world we are all co-producing, knowingly and admittedly or not. For universities willing to genuinely engage and not just play with it, the SDG agenda presents a valuable tool and opportunity, one that requires that they face the weaknesses it exposes in their existing ideas and approaches and help overcome biases towards Level 1 reality and first-order learning. In the following chapter we explore these ideas further within the context of the role of the SDGs in re-thinking research innovation, impact and engagement.

Notes

1. Rickards, L., Steele, W., Kokshagina, O. and Moraes, O. (2020) *Research Impact as Ethos*. RMIT University, Melbourne. https://cur.org.au/project/rethinking-research-impact/
2. Ibid.
3. Godin, B. (2018) *The idea of innovation*. www.internationalinnovation.com See also Godin, B. (2015) *Innovation contested: The idea of innovation over the centuries*. Routledge.
4. Ibid.
5. Marx, L. (2010) Technology: The emergence of a hazardous concept. *Technology and culture* 51, 561–577.
6. Jørgensen, F. A. and D. Jørgensen (2016). "The Anthropocene as a history of technology: Welcome to the Anthropocene: The earth in our hands, Deutsches museum, Munich." *Technology and Culture* **57**(1): 231–237.
7. Ibid., P. 561.

8. Beck, U., Giddens, A., Lash, S. (1994) *Reflexive modernisation*. Polity.
9. Jasanoff, S. (2016) *The ethics of invention: technology and the human future*. WW Norton & Company. P. 247.
10. Jasanoff (2016), p. 19.
11. Funtowicz, S., Ravetz, J.K., (2018) *Post-normal science, Companion to Environmental Studies*. Routledge in association with GSE Research, pp. 443–447.
12. Stirling, A. (2003) Risk, uncertainty and precaution: some instrumental implications from the social sciences. *Negotiating environmental change*, 33–76.
13. Stirling, A., (2009) Participation, precaution and reflexive governance for sustainable development, in: Adger, W.N., Jordan, A. (Eds.), *Governing Sustainability* Cambridge University Press, Cambridge, pp. 193–225.
14. See Jasanoff (2016), pp. 22–23.
15. Schön, D.A. (1983) *The reflective practitioner: How professionals think in action*. Basic Books.
16. Bonneuil, C. and Fressoz, J.P. (2015) *The Shock of the Anthropocene: The earth, history and us*. Verso, London.
17. Plumwood (2013) *Environmental Culture*. Routledge.
18. Allenby, B. and Sarewitz, D. (2011) *The Techno-Human Condition*. MIT Press, Massachusetts.
19. Mejlgaard, N., Bouter, L.M., Gaskell, G., Kavouras, P., Allum, N., Bendtsen, A.-K., Charitidis, C.A., Claesen, N., Dierickx, K., Domaradzka, A. (2020) Research integrity: nine ways to move from talk to walk. *Nature*. 586(7829): p. 358–360.
20. Winner, L. (1978) *Autonomous Technology: Technics-out-of-control as a theme in political thought*. MIT Press. p. 228.
21. Allenby and Sarewitz (2011), p. 44.
22. Accenture 2019 white paper, accessed on http://www3.weforum.org/docs/WEF_Shaping_the_Sustainability_Production_Systems.pdf
23. Making the Fourth Industrial Revolution count for Sustainable Development in Asia and the Pacific, https://medium.com/@UNDPasiapac/making-the-fourth-industrial-revolution-count-for-sustainable-development-in-asia-and-the-pacific-22c351ddd4bf
24. The United Nations Development Programme 2018. *Development 4.0: Opportunities and Challenges for Accelerating Progress towards the Sustainable Development Goals in Asia and the Pacific*. http://www.asia-

pacific.undp.org/content/rbap/en/home/library/sustainable-development/Asia-Pacific-Development-40.html
25. Jasanoff (2016), p. 266.
26. WEF April 2019 White paper *Globalization 4.0—Shaping a New Global Architecture in the Age of the Fourth Industrial Revolution* http://www3.weforum.org/docs/WEF_Globalization_4.0_Call_for_Engagement.pdf Italics added
27. See http://www3.weforum.org/docs/WEF_Dialogue_Series_on_New_Economic_and_Social_Frontiers.pdf
28. Allenby and Sarewitz (2011) p. 184.
29. Jasanoff (2016), p. 267, 266.
30. Woods, N. (2016) Unsustainable development goals? Project Syndicate blog. April 14, 2016. https://www.project-syndicate.org/commentary/unsustainable-development-goals-by-ngaire-woods-2016-04
31. De Saille, S. (2015) Innovating innovation policy: the emergence of 'Responsible Research and Innovation'. *Journal of Responsible Innovation* 2, 152–168.
32. Owen, R., Macnaghten, P., Stilgoe, J. (2012) Responsible research and innovation: From science in society to science for society, with society. *Science and Public Policy* 39, 751–760.
33. Op cit.
34. Ibid., p. 751, 757.
35. Op cit.
36. Goddard, J. (2018). The civic university and the city. *Geographies of the University,* Springer, Cham: 355–373.
37. Wittrock, C., Forsberg, E.-M., Pols, A., Macnaghten, P., Ludwig, D. (2021) *Implementing Responsible Research and Innovation: Organisational and national conditions.* Springer Nature, Berlin. P. ix.
38. Imaz, O., Eizagirre, A. (2020) Responsible Innovation for Sustainable Development Goals in Business: An Agenda for Cooperative Firms. *Sustainability* 12, 6948. Lehoux, P., Pacifico Silva, H., Pozelli Sabio, R., Roncarolo, F. (2018) The Unexplored Contribution of Responsible Innovation in Health to Sustainable Development Goals. *Sustainability* 10, 4015.
39. Jenkins, K.E., Spruit, S., Milchram, C., Höffken, J., Taebi, B. (2020) Synthesizing value sensitive design, responsible research and innovation, and energy justice: A conceptual review. *Energy research & social science* 69, 101727.

40. Jazeel, T. (2019) *Postcolonialism*. Routledge, London. p. 211.
41. Croissant, J.L. (2015) Routine, Scale, and Inequality: Introduction to the Special Issue on Ethics, Organizations, and Science. *Science, Technology, & Human Values* 40, 167–175.
42. Croissant, J.L. (2015) Routine, Scale, and Inequality: Introduction to the Special Issue on Ethics, Organizations, and Science. *Science, Technology, & Human Values* 40, 167–175.
43. Stengers, Isabelle (2018) *Another Science is Possible: A manifesto for slow science*. Polity, London. p. 154.
44. See, for example, Bulkeley, H., Carmin, J., Castán Broto, V., Edwards, G.A.S., Fuller, S. (2013) Climate justice and global cities: Mapping the emerging discourses. *Global environmental change* 23, 914–925.
45. Allenby and Sarewitz (2011) p. 164.
46. Petts, J., Owens, S., Bulkeley, H. (2008) Crossing boundaries: Interdisciplinarity in the context of urban environments. *Geoforum* 39, 593–601. p. 600.
47. For discussion of epistemologies in interdisciplinary research, see, for example, Phoenix, C., Osborne, N.J., Redshaw, C., Moran, R., Stahl-Timmins, W., Depledge, M.H., Fleming, L.E., Wheeler, B.W. (2013) Paradigmatic approaches to studying environment and human health: (Forgotten) implications for interdisciplinary research. *Environmental Science & Policy* 25, 218–228. Murphy, B.L. (2011) From interdisciplinary to inter-epistemological approaches: Confronting the challenges of integrated climate change research. *Canadian Geographer-Geographe Canadien* 55, 490–509.
48. Rickards, L., (2012) *The Melbourne Interdisciplinary Collaboration Exploration Project*. Report to the Melbourne Research Office, University of Melbourne, Melbourne.
49. Ibid.
50. Rickards, L., (2014) *Interdisciplinary Collaboration in Context: Academics and agendas*. University of Melbourne, Melbourne.
51. Meadows, D. (1999) *Leverage Points: places to intervene in a system*. The Sustainability Institute, http://www.sustainer.org/pubs/Leverage_Points.pdf
52. Barry, A., Born, G., Weszkalnys, G. (2008) Logics of interdisciplinarity. *Economy and Society* 37, 20–49.
53. Habermas, Jurgen (1973) *Theory and Practice*, Boston: Beacon Press. p. 1.
54. Plumwood, V. (2002) *Environmental Culture*. Routledge, London. p. 176.
55. Gibson-Graham, J.K. and G. Roelvink (2010). An economic ethics for the Anthropocene. *Antipode* 41(s1): 320–346. p. 324.

56. Ibid., p. 331.
57. Fry, T. (2014) *Cities for a Future Climate*. Routledge.
58. Meadows (1999), p. 16.
59. Roseneil, S. (2011). "Criticality, Not Paranoia: A Generative Register for Feminist Social Research." *NORA—Nordic Journal of Feminist and Gender Research* 19(2): 124–131. P. 129.
60. Roelvink, G. (2016). *Building Dignified Worlds: Geographies of Collective Action,* University of Minnesota Press. p. 164.
61. Roelvink (2016), p. 160.
62. Calisto Friant, M. and J. Langmore (2015). "The buen vivir: a policy to survive the Anthropocene?" *Global Policy* **6**(1): 64–71. Gudynas, E. (2011). "Buen Vivir: today's tomorrow." *Development* 54(4): 441–447. https://www.cnrs-univ-arizona.net/menu-en/covid-am/ecuadors-tragedy-from-buen-vivir-to-dying-badly/
63. Brown, E. and T. McCowan (2018). "Buen vivir: reimagining education and shifting paradigms." *Compare: A Journal of Comparative and International Education* 48(2): p. 317–323. Weber, S. M. and M. A. Tascón (2020). Pachamama—La Universidad del 'Buen Vivir': A First Nations Sustainability University in Latin America. *Universities as Living Labs for Sustainable Development*, Springer: p. 849–862.
64. van Norren, D. E. (2020). The Sustainable Development Goals viewed through Gross National Happiness, Ubuntu, and Buen Vivir. *International Environmental Agreements: Politics, Law and Economics* 20(3): p. 431–458.

5

Re-thinking Research Engagement

The Research-SDGs Relationship

One of the ways in which universities are imagined as enablers to SDGs progress is via their research efforts. Such efforts, however, are also one of the key reasons that universities need themselves to be targets for SDG action. In this chapter we explore the topic of research in order to debunk some of the limited ways in which research is positioned within conventional discussions about the SDGs and underline how the SDG agenda is a heterogeneous mix of imaginaries and intentions, whose actual manifestation in the world depends in large part on how groups such as researchers engage with it.

In contrast to the prevailing idea that the SDGs are a potential research topic that researchers are free to contribute to (or ignore) if they wish—or the more specific idea that the SDGs are merely a tick-a-box list of themes that can be mapped superficially to existing research efforts—and in contrast to the argument that claims that the SDGs as a top-down imposition on academic freedom, we offer a critical and holistic perspective on the research-SDG relationship. It is one attuned to the historical role of research in generating the unsustainable state we are currently in; the fact that research is a physical and social activity that generates impacts continuously (not just at the end when researchers are 'ready'); and the very real ways in which research is profoundly reliant on the success of the SDGs for its long-term future.

The research-SDGs relationship includes their shared underlying belief in development, broadly defined. As a normalised social activity, research is development-like not only because it is, for better or worse, directly or indirectly, a product and driver of social and economic development, but because at base it is a faith-filled intervention to help unfold imagined potential. Thinking about sustainable development exposes the way that, despite this resonance, research has failed to keep up with the stinging critiques of development's social effects and processes, and the emergence of sustainable development to address environmental issues and the existential threat they present.

Although research is development-like, but because it has positioned itself as a purported distant observer and often disguised itself as a mere processor of others' values and wishes, it has not been subject to the sort of fierce reflexivity and renovations that social and economic development have. As a development process, research now urgently needs to become more *like* sustainable development if it is to contribute usefully to sustainable development. Regardless of topic area, discipline or institution, research needs to become more aware of complexity, uncertainty and the deeply political and ethical nature of all research endeavours (including those endeavours that are conspicuous in their absence).

Combined with the practical challenges that the SDGs pose to research, such as calling out its inequalities, resource consumption and greenhouse gas emissions, this means that intellectually, ethically and practically the SDGs are a wake-up call for university research as much as the latter is an instrument for achieving the SDGs. Indeed, as Flurina Schneider and colleagues argue, 'transforming society towards sustainable development *requires* higher education institutions to transform science and research itself more broadly'.[1]

Centrifugal Forces

University research and scholarship has classically been imagined as mental work, not just in the sense that it takes a lot of thought, but in the sense that is seen as qualitatively and spatially distinct from the world and its bodies as we discussed in Chap. 3 on the role of the University. Encapsulated in the distinction that French philosopher and mathematician Rene Descartes famously articulated between the Mind and the Body, universities—and scientific research and humanities scholarship in particular—have become symbols of the human Mind. Given that possession of a Mind is taken as the basis of our species' purported superiority in the world, academia's association with it has predominantly been a sign of status. It is an association academia has e worked hard to protect, including ongoing efforts to maintain apparent independence from government, economic interests and the 'sullying' effects of politics.

It also includes ongoing efforts by researchers to maintain apparent independence from their bodies, emotions and context, as feminist, postcolonial and Indigenous scholars have long critiqued. Enabling and contributing to this imagined 'hyper-separation' of researchers from their context, as Val Plumwood might describe it,[2] have been two developments within academic research: one towards abstraction and one towards specialisation. Both continue to pull researchers out of messy empirical research contexts, the first towards representations, theories and claims of universal knowledge, the second towards narrow, disciplinary debates about an ever-thinner slice of reality.[3] Together these centrifugal forces have turned researchers inwards towards academia and its tangle of networks, languages and spaces, including self-serving and self-served evaluations of research quality and ever louder claims to universality.

The symbolic divide between academic research and the world has encouraged a basic physical one. Regardless of whether our research outputs place us at the top of the academic knowledge hierarchy in which contributions to universal theories rule, many of us inhabit the same sort of 'carpet worlds' as the contemporary and colonial office-based bureaucrats as anthropologist Tess Lea describes in her analysis of policy makers governing the lives and landscapes of Indigenous Australians. Like policy makers, many academics spend:

> the majority of the their waking hours inside spaces with controlled temperatures and recycled air, facing monitors nested on laminated desks or held in hands, kitchens with dictatorial rules about the treatment of dirty dishes and fridge waste, and emergency stairwells that repel any but vermin and microbial life from their bare concrete and stale aromas.[4]

Despite this, all of us are nevertheless 'relationally enmeshed with other substances and circulations, microbial, local and global',[5] including the ones energising and emanating from the computers on our desks, and the capital flows energising our institutions.

We are also enmeshed with competing socio-technical imaginaries and shifting social expectations and demands. Just like viruses and electricity, our relationship with these 'exogenous forces' cannot be externalised forever. As Raewyn Connell argues, a Good University is sustainable, where sustainable means not only an absence of negative environmental effects but the more general 'capacity to flourish as an organization over the long run'. As she notes:

> A university is not a pop-up shop: its work needs time to unfold. That needs a steady source of income. … It needs resilience in the face of disruption, change and political pressure.[6]

Arguably the most significant outside pressure upon universities—the one that has already disrupted its income flows significantly, adding immediate pressure—is the very need to pay more attention to the 'outside world'. Governments, businesses and communities, we are told, are sick of academics being hunched over their desks, hidden behind university fences, talking among ourselves. It is time to go outside. Or more accurately, it is time to acknowledge that we, like everyone else, are never not in and of the world. This includes universities as we described in Chap. 3.

Centripetal Forces: The Great Acceleration

One of the most pressing reasons to attend to 'worldly matters' is the question of funding. Obtaining funding for research without giving up the autonomy to determine the direction and content of research has been a long-standing conundrum for academia. In the US, prior to the World Wars university research was predominantly funded by philanthropies, corporations and wealthy individuals. Although agricultural research was supported through government-funded Land Grant Universities, most research remained staunchly independent of government support. Underlying this policy was the spectre of the 'monster' of Lysenkoism: a fear of government intervention in research. As the name suggests, this fear was stimulated by the Soviet regime's control and perversion of science in the 1930s and 1940s, including its execution of geneticists whose views on agriculture conflicted with the regime's favoured Lamarckian agricultural science theories.[7]

While the US had a proudly different system to the Soviet Union, one that it felt showcased the fruits of capitalism, during the war university science proved exceedingly useful for national interests. The war also interrupted and undermined the supply of new research findings from Europe and its colonial networks. As the war ended, a new era of more nationally oriented university research emerged in the US. The same occurred in Europe where universities were also struggling to re-establish and reposition themselves. In the UK, the Committee of Vice-Chancellors wrote to the UK government in 1946, declaring that:

> In the view of the Vice-Chancellors ... the universities may properly be expected not only individually to make proper use of the resources entrusted to them, but collectively to devise and execute policies calculated to serve the national interest. And in that task, both individually and collectively, they will be glad to have a greater measure of guidance from the Government than, until quite recent days, they have been accustomed to receive.[8]

They were successful in convincing the government to largely take over the cost of expanding the universities, and while the universities generally dodged paying their offering price of providing 'more direct support for the national interest', the idea of tying university research to national development was born.[9] In the US, the federal government was persuaded to similarly expand funding for university research by Vannevar Bush, author of *Science—The Endless Frontier*. He argued for academic-led 'basic' research on instrumental grounds, asserting that university research was part of the division of labour and flow of knowledge needed to lead the broad project of post-war reconstruction and social and economic development.

> Basic research is performed without thought of practical ends. ... The scientist doing basic research may not be at all interested in the practical applications of his work, yet the further progress of industrial development would eventually stagnate if basic scientific research were long neglected. ... We can no longer count on ravaged Europe as a source of fundamental knowledge. ... In the future we must pay increased attention to discovering this knowledge for ourselves particularly since the scientific applications of the future will be more than ever dependent upon such basic knowledge. ... New impetus must be given to research in our country. Such impetus can come promptly only from the Government. ... [W]e cannot expect industry adequately to fill the gap.[10]

Underlining that 'the social sciences, humanities, and other studies so essential to national well-being' had to part of the expansion of national research capacity, Bush stressed that independence and patience was needed:

> Support of basic research in the public and private colleges, universities, and research institutes must leave the internal control of policy, personnel, and the method and scope of the research to the institutions themselves. This is of the utmost importance. ... Basic research is a long-term process—it ceases to be basic if immediate results are expected on short-term support.[11]

In the decades that followed, US research capacity expanded rapidly. While about half of this was in military research to 'prepare for the next war' (and by 2012 the US was spending approximately four times the amount of all other countries in the OECD combined on Defence research and development (R&D)[12]), 'nondefence R&D' also increased dramatically.[13] Although the majority of this went to outer space research

(NASA) or medical research,[14] much of it was to drive socioeconomic development.

US and European post-war activities helped drive what Anthropocene scientists now call the 'Great Acceleration'—as discussed in Chaps. 2 and 4 this is the post-1945 period in which material production and consumption, and attendant planetary degradation and destabilisation, have skyrocketed exponentially.[15] An 'eccentric historical moment', 'the most anomalous and unrepresentative period in the 200,000-year-long history of relations between our species and the biosphere',[16] the Great Acceleration can be considered the third stage of the Anthropocene, after the profound alteration of the planet kick started by colonialism and the industrial revolution. Though third, the post-war period is favoured as the official start date of the Anthropocene due to it being when 'the most rapid and pervasive shift in the human-environment relationship began'—a 'dramatic change in the magnitude and rate of the human imprint' on the planet[17] that surprised scientists when they mapped changes from 1750 onwards.[18]

As many have pointed out, and the whole question of development addresses, global aggregates hide enormous inequities. The *Anthropos* of the Anthropocene is far from humanity as a whole, and in fact countless humans have not only *not* caused the Anthropocene but have been killed in the violent and exploitative colonial and industrial processes that have driven it, while many more are now dying indirectly because of it.[19] When disaggregated, the Great Acceleration graphs depict the development perversities involved:

> In 2010 the OECD countries accounted for 74% of global GDP but only 18% of the global population. Insofar as the imprint on the Earth System scales with consumption, most of the human imprint on the Earth System is coming from the OECD world.[20]

Further disaggregation would demonstrate great inequities within nations as well.[21] All scales—global, national, subnational—illustrate the immense problem that the SDG agenda is trying to address.

Within the US and other OECD countries, the Great Acceleration has been driven by a huge ramping up of industry and research capacity, repurposing the immense leap in capacity generated for the war-time effort. Unsurprisingly, as in the war, much of this went hand in hand as academia's inherent centrifugal forces have been countered by the strong magnetic force of funding priorities. Despite the turn to government-funded research in the US and other countries such as Australia, the Great Acceleration period is one in which university research has been increasingly harnessed to industry in direct and indirect ways within many countries. While most national governments have avoided intimate control of university research, university research has become a government policy issue and tool and used to funnel research towards select development ends, giving research a centrifugal quality and scattering its impact far afield.

Not long after government funding of university research was introduced, so too was the idea of 'oriented' or 'use inspired' research and national research priorities.[22] Although some research has been oriented towards global and public good issues (as discussed further below), in many contexts this reorientation fostered an institutional logic of what Sá and Sabzalieva call 'scientific nationalism'. Focused on using research to help generate *innovation* (i.e. practical outcomes, not just research), the rationale for scientific nationalism is largely economic development.[23] It is an economic mindset that has in turn led to a valuation of research in economic terms, with universities under pressure to demonstrate they offer a good return on investment (ROI).[24]

Whether by pushing universities towards industry partners and 'innovation systems' and/or simply directing a greater or lesser degree of its own funding towards industry and economy-oriented impacts, government policies have generally fostered the idea that the value of research ought to be measured in terms of traceable, tangible contributions to national economies more than to international public goods, with the exception of high-profile, blue-sky scientific discoveries.[25]

Although nationalistic research priorities are often justified on the basis of allaying taxpayers' fears about government spending on university research, social research indicates that public trust in university research is declining. But rather than declining because of a perception university research is too 'divorced from the real world' (as governments assume), public trust is generally withering because of dis-ease about sciences' ambiguous relations with big business. For example, in nutrition science—a vital area of research for addressing SDG 2 on hunger and malnutrition and SDG 3 on good health and wellbeing, among many others, but predominantly focused on industrial world food and beverage consumption—there is growing concern that the field is not perceived by the public as trustworthy and credible.[26]

Discernible here is not Lysenkoism but its equally feared monster twin, 'sponsorism': the corrosive effects of corporate influence on university research. It is a fear that is not ill-founded. Even where academic research has largely avoided direct corporate sponsorship out of concern about the perceived conflict of interest, the effects of corporate interest are arguably detectable, thanks in part to their champions in government.

One of the ironies for science is that its fierce commitment to being 'free' of politics and instead seizing the tenets of positivism means that it is poorly equipped to recognise, reflect on and react to the way some research agendas have been prioritised over others during the Great Acceleration. The very topic of ethics is dismissed as political and irrelevant, a matter of philosophy, personal preference and worldly debate, not scientific inquiry save for the need to deal with the occasional episode of professional misconduct. Hence the need for what we have termed 'ethical innovation' with its emphasis on responsibility, attentive, disruptive, authentic and regenerative (see further detail in Chap. 4).

In agricultural science, for instance—an applied field that still has to work strenuously to prove itself as a hard, disembodied scientific discipline with a legitimate place within the university system—its evolution into a science (strongly focused on genetics, not Lamarckianism) has gone hand in hand with the standardisation, professionalisation and industrialisation of agriculture. As a practice, this took place first in wealthy nations and then in the dozens of other nations to which the European and then American agricultural model has been exported via colonialism and international development.[27] This striking homogenisation of agricultural knowledges and practices, as well as animal bodies, plants and landscapes, is largely unremarked upon within the discipline, however, because such political influences and processes are externalised as irrelevant.

Ethicist Paul Thompson suggests that while such questions regularly animate informal discussions within the field, they are systematically excluded from actual scientific discourse, cast as something more suited to the staff kitchen than a paper's conclusion:

> Where are the books and articles in which the scientists of the 1950s and 1960s articulated the rationale for developing chemical pesticides, herbicides, and fertilizers? Where are the course syllabi in which instructors in the agricultural sciences discussed alternative approaches for understanding agriculture's impact on the broader environment? ... The failure to record the considerations and deliberations that led scientists to undertake the studies that led to the rise of chemical and molecular technologies in the plant sciences, and to a mechanical revolution in animal husbandry, has left the current generation vulnerable to the charge that such developments were undertaken in secret by profit- and power-seeking individuals with little regard for farmers, farm animals, the environment, or the broader public. Even under assault from authors such as Rachel Carson or Wendell Berry, the agricultural disciplines of the 1960s, 1970s, and 1980s displayed too little willingness to articulate the reasons for, values behind, and logic of their science.[28]

Rachel Carson was among those who first called out the systematic influence of corporate agribusiness on agricultural science. Denigrated by industry and government officials after the publication of *Silent Spring* as a 'hysterical woman' (emotional body), she used every speaking engagement she had to ask the hard questions: 'When a scientific organization speaks, whose voice do we hear, that of science or of the sustaining industry?'[29] Thompson takes up these questions, asking:

> Does the current generation of scientists see no reason to articulate the rationale for doing what they do, to engage in self-reflection, or for defending what they do against mounting criticism?[30]

In general, the answer is mere silence. The neoliberalisation that has favoured industrial agriculture has also heightened the atmosphere of individualistic competition within academia, leaving little time, energy or inclination to worry about, little less debate, 'what it all means'. In the never-ending race to secure grants, big picture anxieties are seen as unrealistic if not unscientific. And in all academic areas, not just agricultural science or science in general, the drive to abstraction and specialisation mentioned above blinds us to the context we are part of, or at least the role of our research within it. We are too busy to look outside the window long.

Conscience and Excellence

Somewhat militating against this myopia is another research-society model that equally emerged within the Great Acceleration and features to a greater or lesser degree within different nations today, though in the shadow of an increasingly towering scientific nationalism. Referred to by Creso Sá and Emma Sabzalieva as 'scientific globalism', this logic or imaginary presents science as 'a global endeavour with norms deriving from the scientific community, cutting across political, ethnic and cultural borders ... not a means to an end but a quest for discovery, oriented towards universalist ideas such as the betterment of human society'.[31] Increasingly multidisciplinary, this is the science of global challenges, international consortia and World Class Universities. Whether oriented towards advancing human understanding, or bettering human society more directly, scientific globalism is part of the framework in which the SDG agenda has emerged and is similarly entwined with the rise of sustainable development, though as we discuss it also poses barriers to effective SDG engagement.

While strongly oriented to basic research and blue-sky discoveries, scientific globalism also celebrates research's capacity to help address 'global challenges' such as climate change, pandemics and cybersecurity. Combined with the humanistic ideal of knowledge for and defining of humanity, scientific globalism is associated with public good research. Compared to the more directly economistic focus of scientific nationalism, scientific globalism helped encourage the rise of ecology and other environmental sciences in the post-war period. Fairfield Osborn's *Our Plundered Planet* and Willian Vogt's *Road to Survival*, both published in 1948, helped draw attention to the Earth as a planet in peril. Like the

Club of Rome's Limits to Growth report thirty years later, Osborn predicted that unchecked consumption and human population would lead to disaster, while Vogt calling instead for an 'ecological approach'.[32]

By the 1970s, ecology was, like the emerging field of climate change science, a key 'impact science' calling into question the un-reflexive character of the 'production sciences' (e.g. geology, chemical engineering) and their aggressive application in the name of economic development.[33] Paul Sears suggested that ecology became in fact a 'subversive science', challenging assumptions and practices of modern societies being, therefore, of universal relevance.[34]

Encouraging the impact sciences' claims to universal relevance has been an expansion in their focal scale from plants in local fields to the planet as a whole. The gradual scientific discovery and elaboration of human-induced global climate change has been enabled by a 'vast machine' of transnational scientific collaborations and infrastructure.[35] Helping establish this machine was the same step change in computing power that drove other aspects of the Great Acceleration including the Manhattan Project—the world's largest ever scientific project at the time.[36] It is the Manhattan Project that triggered the US government's U-turn on investment in science, paving the way for state-funded university research and the subsequent industrial development.

The project also triggered a cascade of atomic bombs, including an estimated 2000 detonated around the world in the name of research, all of which left behind waves of death and destruction. Also left behind were trails of radioactive materials whose ongoing circulations through the planet's water, ice, air, land and living subsystems converted the Earth into a giant laboratory[37] and informed the science that now realises that the Earth operates as a single system, albeit one we are rendering dysfunctional. It is no coincidence that not only is the post-war 'Nuclear Age' and longer Great Acceleration considered to be the main Anthropocene period to date, but the sharp line of radioactivity injected into the Earth's strata with the detonation of the Manhattan Project's first bomb at 5.29 am on July 16, 1945, is the favoured candidate for the epoch's geological marker.[38]

Though he had no idea what significance its bombs would have in subsequent science, politics or planetary history, Robert Oppenheimer, leader of the Manhattan Project, did have a belated sense of the human horror his work helped unleash, as indicated in Arendt's characterisation of him in Chap. 3. He named the first bomb the Trinity Bomb after the Holy Trinity in recognition of his God-like position and recited to himself after the detonation, 'Now I am become Death, the destroyer of worlds'. In science around the world, the project and war triggered a 'crisis of conscience'[39] 'as the social consequences of scientific discovery and collaboration became readily apparent'.[40]

Among the impacts was a renewed effort to distance science from society and even designate science as independent of society *by definition*. During this period, Robert Merton influentially argued that science should be distinguished by a commitment to four cultural values: universalism (separating scientific claims from those who make them, such that their personal characteristics are irrelevant); communalism (scientific knowledge as communal to humankind, and not something that should be privatised); disinterested (not distracted by others' interests or opinions, including funders); and scepticism (always open to questioning 'truths' within their disciplines).[41]

Merton's scientific ideals continue to help define academic research today, especially as scientific nationalism intensifies and with it the need to defend science's independence. The ideals are especially apparent in the contemporary notion of 'research excellence', which is arguably the most influential imaginary shaping academia today. As Erika Kraemer-Mbula and colleagues explain:

> Perceptions of what constitutes 'good science' shape the progress of knowledge creation and knowledge-based innovation. Globally, 'good science' affects decisions about what is funded, and what is not. It dictates who is rewarded and encouraged to pursue research. It promotes certain disciplinary traditions, but likewise discounts and discourages others. However, in the ever-competitive world of science and research, 'good' may not be good enough anymore. 'Excellent' science and associated prestige is increasingly seen as more valuable—something one should strive for. Not surprisingly, 'excellence' has become a buzzword, more popular than the underlying core notion of 'quality'. Those who are seen to be producing 'scientific excellence' are elevated to the highest paid jobs in the most prestigious

institutions, granted greater degrees of academic leeway and expression, lauded as 'thought leaders' by peers, and turned to for policy and practice insights in the non-scientific realm. What gets called excellent, steers and influences the behaviour of individual researchers and teams, research organisations and research funders, and affects society at large.[42]

The ideal of research excellence is coupled with that of academic autonomy via the assertion that it is only unencumbered academics who can ascertain if research is excellent. The relationship is laid out in Michael Polanyi's influential 1962 *The Republic of Science*. While acknowledging that 'Emergencies may arise in which all scientists willingly apply their gifts to tasks of public interest', he argued that '[a]ny attempt at guiding scientific research towards a purpose other than its own is an attempt to deflect if from the advancement of science'.[43] For Polanyi:

> The Republic of Science is a Society of Explorers. Such a society strives towards an unknown future, which it believes to be accessible and worth achieving. In the case of scientists, the explorers strive towards a hidden reality, for the sake of intellectual satisfaction. And as they satisfy themselves, they enlighten all men and are thus helping society to fulfil its obligation towards intellectual self-improvement. A free society may be seen to be bent in its entirety on exploring self-improvement—every kind of self-improvement.

In this vision, the self-improvement of human society is the goal; science and thus universities are a key route to it. Science and society are kept on the path to improvement and freedom thanks to the self-organising discipline and authority of science as a whole. This authority stems from individuals fastidiously exercising their disciplinary expertise to uphold standards and promote scientific values (plausibility, accuracy, systematic importance, intrinsic interest and originality) in their judgement of others' work. Continuing the imperial theme of exploration: 'The more widely the republic of science extends over the globe … the more clearly emerges the need for a strong and effective scientific authority to reign over this republic'.[44]

Crucially, for this republic to function effectively, scientists need protection not only from 'the interference of political or religious authorities', but from 'corrupting intrusions and distractions'. It is this need for protection that justifies the location of science within universities as special places of scholarship. As Polanyi outlined:

> For though scientific discoveries eventually diffuse into all people's thinking, the general public cannot participate in the intellectual milieu in which discoveries are made. Discovery comes only to a mind immersed in its pursuit. For such work the scientist needs a secluded place among likeminded colleagues who keenly share his aims and sharply control his performances. The soil of academic science must be exterritorial in order to secure its control by scientific opinion.[45]

In other words, the Republic of Science is imagined as encompassing the globe, but not being of it; as examining and leading the world, but existing separate to it.

The value of scientific independence and rigour goes without saying and is something we unpack and argue for further below. It is important to note, however, the double move that this influential framing of scientific truth performs: simultaneously positioning scientific knowledge as, on the one hand, universal, aspatial, apolitical and definitive of human improvement, and on the other hand, locating it within the discerning minds of an exceptional, privileged and gendered group of individuals, 'secluded' and confined to 'like-minded colleagues' within the unique space of universities, free from distractions such as the ultimate uses of their discoveries or the state of the world that sustains them.

Re-thinking Freedom

The Republic of Science and its ideals are now manifest in a large international network of universities, ranked multiple times per year according to various metrics that bestow prestige and market advantage. While the idea of research excellence is at the heart of these developments, the relation of universities to each other and to society has been complicated by the amplification of two complementary agendas that complement and challenge the simplicity of an Academy devoted purely to excellence. The first of these is 'research impact', associated with the concerns outlined above about academic research's intended and unintended effects in the world. As David Cash and colleagues outline in a seminal paper on 'Knowledge systems for sustainable development', researchers now need to produce research that is not only rigorous ('excellent') but also relevant (salient, timely, usable and effective) and legitimate (endorsed by 'end users' and other stakeholders).[46]

The 'research productivity' agenda is somewhat similar, but the focus is on research outputs as an end in themselves. As the name suggests, the aim is to increase the magnitude of academic knowledge and the efficiency of its production. If the research excellence agenda is about quality, the research productivity agenda is about quantity, preferably in prestigious outputs. It takes to heart the assertion we are in a knowledge economy and seeks to grow knowledge, not to expand human understanding as much as to harness it as the commodity it has become.

All academics are now well accustomed to the pressure of producing ever more academic publications (notably in 'highly ranked' journals) and 'bringing in' ever more research income as a measure of productivity and thus worth in and of itself. Despite enormous differences in measures of quality and the norms of writing and publishing between the STEM (science, technology, engineering and mathematics) and HASS (humanities, arts and social sciences) disciplinary areas, the productivity agenda is largely one-dimensional and science-centric.

Academic knowledge production is made commensurate across all areas of the university, enabling its conversion into dollar values to enable, in turn, cost-benefit analyses, rankings and the universal application of a capitalist 'growth' mentality. Even if left free to determine their own particular direction, universities have been harnessed tightly to the capitalist growth imperative. Rather than simply being expected (by some governments) to provide an external input to economic development, university research has become animated by it from within.

This is doubly so when it is recognised that the 'common sense' imaginary of economic growth originated in part within university departments. While it is beyond the scope of this chapter to discuss the evolution of economic thought, it is important to appreciate that universities' absorption of the economic growth doctrine further illustrates the role of universities as both catalysts and targets for development. As Giorgos Kallis points out, economic growth is a recent phenomenon of industrial capitalism. Economists first measured it in the 1930s and its pursuit become universal only from the 1950s.[47]

This brings us back to post-war Anthropocene America. Key among the drivers of the new growth agenda was the Paley Commission's 1952 report *Resources for Freedom: Foundations for Growth and Security*, authored by representatives from the fossil fuel industry and economists from Harvard University and Aubrey College. Through the far-reaching impact of its narrative and ideas, it helped catalyse the Anthropocene that universities around the world are now caught within. Among its powerful arguments was that the 'the economy' is an 'apparently tangible, discrete object'[48] that ought—like universities—be unshackled from 'the dead hand of government'.[49] It addressed the threat of resource scarcity highlighted in Osborn's *Our Plundered Planet* and Vogt's *Road to Survival* (mentioned above) as well as the Club of Rome's *Limits to Growth* report but reformulated such scarcity as an abstract question of price and a challenge to scientists and others to develop technological substitutes. It cemented the idea, institutionalised in the post-war Bretton Woods agreement, that national economies could be measured and ranked by a Gross National Product calculation and that such a measure should not be limited to a nation's 'residential unit' but include its off-shore activities. It encouraged such off-shore activities by displacing the idea of a US self-sufficient in resources with the vision of a nation controlling an expansive, international web of supply chains[50] in order to expand its geopolitical power and outsource its dirty production processes, leaving the nation to focus on capitalist consumption and the cerebral work of the knowledge economy and its increasing financialisation.

This contribution to global development pathways was bolstered by further economic theories, including Walt Whitman Rostow's 1961 *Stages of Economic Growth: A non-communist manifesto*, which naturalised the US model as the epitome of socioeconomic development, an imagined telos of 'unlimited production' and naturalised, normalised growth.[51] As part of the normalisation of growth, the theory now animates universities and their escalating competition in the knowledge economy. Despite being the intellectual product of only one particular arena of academia, economic growth is now a hegemonic (dominant and accepted) idea in society across and beyond universities, what the late Italian theorist Antonio Gramsci would have called a 'common sense'.

Gramsci would have appreciated how universities have helped such an idea become hegemonic, not only by helping generate it in the first place, or by implementing it at a sectoral as well as organisational level (structuring the whole productivity agenda for higher education around it) but being part of the 'civil society' that implicitly endorses it through its lack of resistance. At the same time, he would have appreciated that universities do house counter-hegemonic ideas—such as the increasingly respected idea of degrowth that Giorgos Kallis and others are working hard to advance, despite the still marginalised status of ecological economics within the Academy[52]—and are not intrinsically limited to reproducing such self-harming policies.

Gramsci's understanding of the diffuse and intimate interlinkages between the state and society helps to throw into question the ideal of academic freedom expressed in Polanyi's Republic of Science and the vision of a free Economy that neoliberal economists especially advocate for. Like the Cartesian conceptions of Human and Nature on which they are based, the two are not as separate as they seem. As Sheila Jasanoff forcefully argues, the idea that science or universities are separate to society is a carefully manufactured construction, a performance that allows for subsequent controlled exchanges (funding here, advice there etc.) between them.

French philosopher Michel Foucault similarly emphasised the false premises of separation on which dominant notions of freedom rely. As he put it, 'there is no pure freedom to be emancipated, just as there is no pure power to dominate it'.[53] In 'Discipline and Punish', he argues that autonomous individuality is an intended effect of the sort of power at work within universities, where individuals are enculturated with a whole range of assumptions and norms. Rather than escaping all such institutions and their influence, Foucault argued that any resistance 'must work through, not merely against, power. This means trading in the model of freedom as autonomy for a more experimental model of freedom … and working through, in, and alongside of power.' He concluded that 'resistance to modern practices of power requires resistance to modern practices of freedom'.[54]

Similarly, Brian Massumi argues that in the face of Capitalism—that machinic imaginary that seeks to control and reproduce what is recognised as 'valuable', even within the research world—we need to cultivate practices of 'creative duplicity'. Such practices recognise our emplacement within Capitalist institutions and landscapes, but nevertheless work to nurture alternatives 'within its pores'.[55] Academia has an advantage here thanks to its 'quasi-sovereign' character,[56] stemming from its origins in the Church and the subsequent coproduction of 'independence' from government and business.

As academics have noted in debates about the financialisation of universities, for example, while aspects of universities such as their substantial real estate portfolios lend themselves to generic financialisation processes (i.e. to being geared in such a way that they generate profit-making opportunities), other aspects are extremely poorly suited to such reconfiguration. Not only does this poor fit lead to incredible damage within universities (e.g. mounting, unpayable student debts, cuts to staff pensions because they sit awkwardly in university ledgers[57]), it is a crucial reminder of academia's 'peculiar' nature[58] and the opportunities for independence of spirit and creativity this represents.

Activating the Potential of the SDGs

For some academics, the SDGs are an imposed, top-down agenda, one that they resist because it affronts their belief in academic freedom and/or because it clashes with their existing commitment to using research for national economic development. Yet there are layers upon layers of irony in these efforts to dismiss the SDG agenda and the Anthropocene condition it draws attention to. The first is simply the point, highlighted above, that all academics are already beholden to imposed agendas, whether the paradigms of their particular disciplines, the productivity and excellence agendas of higher education, capitalist 'research impact' models, or the pervasive and naturalised intellectual structures of the Anthropocene such as never-ending economic growth.

Second, as discussed further below, the SDG agenda encompasses economic growth and is rejected by other academics and activists for this very reason—for being too conservative, not too radical, for being too light of an imposition on social practices and thought, not too difficult an ask. Third, the idea of being able to choose whether one engages with the SDGs and underlying Anthropocene condition is based in the myth of the disembodied academic and Polanyi's 'exterritorial' university. The question is not whether one engages with sustainable development and the future, it is a matter of how.

Finally, attempting to distance academia from the Anthropocene now—after it has helped drive colonialism, industrialisation and the Great Acceleration—is simply too convenient. Instead, we need to look deep into academia and find the points of resistance and seeds of counter-hegemony that have long existed but have been marginalised by the ascent of particular knowledge groups and their imposition of standards and norms across the sector. We need to work not against, or uncritically for, universities and the SDG agenda, but 'through, in and alongside' the power that they offer individually and at their many points of overlap. In the following chapter we turn to focus on the critical role of learning and teaching (L&T) *about, for* and *through* the SDGs as part of a transformative societal agenda.

Notes

1. Schneider, F., Kläy, A., Zimmermann, A.B., Buser, T., Ingalls, M., Messerli, P. (2019) How can science support the 2030 Agenda for Sustainable Development? Four tasks to tackle the normative dimension of sustainability. *Sustainability science* 14, 1593–1604.
2. Plumwood, V. (1993) *Feminism and the Mastery of Nature* Routledge, London.
3. Peter Weingart 2010 in Ox HB on IDR.
4. Lea, T 2020, *Wild Policy*, Stanford, Stanford University Press, p. 27.
5. Ibid.

6. Connell, R (2019) *The Good University: What Universities do and why it's time for radical change*, Melbourne, Monash University Press, p. 174.
7. Dennis (2015) Our monsters, ourselves, in Jasanoff and Kim.
8. Polanyi, M. (1962) The republic of science. *Minerva* 1, 54–73. P. 13.
9. Op cit.
10. Bush, V. (1945) *Science—The Endless Frontier*. A report to the President by Vannevar Bush. Director of the Office of Scientific Research and Development, United States Government Printing Office, Washington.
11. Ibid.
12. Archer, C., Willi, A., (2012) *Opportunity Costs: Military Spending and the UN's Development Agenda. A view from the International Peace Bureau*. International Peace Bureau, Geneva.
13. For a chart of US R&D outlays 1953–2020 see https://www.aaas.org/sites/default/files/2020-05/Function.png
14. See https://www.aaas.org/sites/default/files/2020-05/FunctionNON.png
15. Steffen, W., Broadgate, W., Deutsch, L., Gaffney, O., Ludwig, C. (2015) The trajectory of the Anthropocene: The Great Acceleration. *The Anthropocene Review* 2, 81–98.
16. McNeill, J.R., Engelke, P. (2016) The Great Acceleration. Harvard University Press. P. 5.
17. Steffen, W., Sanderson, A., Tyson, P.D., Jager, J., Matson, P.A., Moore, B., Oldfield, F., et al., (2004) *Global Change and the Earth System: A planet under pressure*. Springer, Berlin.
18. Angus, I. (2016) *Facing the Anthropocene: Fossil Capitalism and the Crisis of the Earth System*. NYU Press.
19. Gemenne, F., (2015) *The Anthropocene and its victims*, in: Hamilton, C., Bonneuil, C., Gemenne, F. (Eds.), The Anthropocene and the Global Environmental Crisis: Rethinking modernity in a new epoch. Routledge, London, pp. 168–174.
20. Steffen, W., Broadgate, W., Deutsch, L., Gaffney, O., Ludwig, C. (2015) The trajectory of the Anthropocene: The Great Acceleration. *The Anthropocene Review* 2, 81–98. P. 91.
21. Angus 2016.
22. Oliveira, M.B.d. (2014) Technology and basic science: the linear model of innovation. *Scientiae Studia* 12, 129–146.
23. Sá, C., Sabzalieva, E., (2018) Scientific nationalism in a globalizing world, in: Cantwell, B., Coates, H., King, R. (Eds.), *Handbook on the Politics of Higher Education*. Elgar, pp. 149–166.

24. Ibid.
25. Godin, B. (2009) National innovation system: The system approach in historical perspective. *Science, Technology, & Human Values* 34, 476–501.
26. See, for example, Garza, C., Stover, P.J., Ohlhorst, S.D., Field, M.S., Steinbrook, R., Rowe, S., Woteki, C., Campbell, E. (2019) Best practices in nutrition science to earn and keep the public's trust. *The American Journal of Clinical Nutrition* 109, 225–243.
27. Rickards, L., (2006) *Capable, Enlightened and Masculine: Constructing English Agriculturalist Ideals in Formal Agricultural Education, 1845–2003*. D.Phil. thesis, School of Geography and Environment. University of Oxford, Oxford.
28. Paul Thompson, Foreword, In Zimdahl, R.L. (2012) *Agriculture's ethical horizon*. Elsevier. Pp. xi–xii.
29. Quoted in Linda Lear, Afterword, in Rachel Carson (2000 [1962]), *Silent Spring*. Penguin.
30. Thompson, in Zimdahl (2012).
31. Sá and Sabzalieva (2018) p. 151.
32. Robertson, T. (2012) Total war and the total environment: Fairfield Osborn, William Vogt, and the birth of global ecology. *Environmental History* 17, 336–364.
33. Schnaiberg, A. (1980) *The Environment: From Surplus to Scarcity*. Oxford University Press., New York.
34. Sears, P.B. (1964) Ecology—a subversive subject. *Bioscience* 14, 11–13.
35. Edwards, P.N. (2010) *A vast machine: Computer models, climate data, and the politics of global warming*. MIT Press.
36. Sá and Sabzalieva (2018).
37. Masco, J., (2015) Terraforming planet Earth, in: DeLoughrey, E., Didur, J., Carrigan, A. (Ed.), *Global Ecologies and the Environmental Humanities: Postcolonial Approaches*. Routledge, London, pp. 307–333.
38. Masco, J. (2013) *The nuclear borderlands: The Manhattan project in post-cold war New Mexico*. Princeton University Press.
    Steffen, W., Leinfelder, R., Zalasiewicz, J., Waters, C.N., Williams, M., Summerhayes, C., Barnosky, A.D., Cearreta, A., Crutzen, P., Edgeworth, M., Ellis, E.C., Fairchild, I.J., Galuszka, A., Grinevald, J., Haywood, A., Ivar do Sul, J., Jeandel, C., McNeill, J.R., Odada, E., Oreskes, N., Revkin, A., Richter, D.d., Syvitski, J., Vidas, D., Wagreich, M., Wing, S.L., Wolfe, A.P., Schellnhuber, H.J. (2016) Stratigraphic

and Earth System approaches to defining the Anthropocene. *Earth's Future* 4, 324–345.
39. Hellström, T., Jacob, M. (2005) Taming unruly science and saving national competitiveness: Discourses on science by Sweden's strategic research bodies. *Science, Technology, & Human Values* 30, 443–467.
40. Sá and Sabzalieva (2018) p. 151.
41. Merton, R.K. (1942). *The Sociology of Science: Theoretical and Empirical Investigations*. Chicago: University of Chicago Press.
42. Kraemer-Mbula, E., Tijssen, R., Wallace, M.L., McClean, R., (2019) Introduction, in: Kraemer-Mbula, E., Tijssen, R., Wallace, M.L., McClean, R. (Eds.), *Transforming Research Excellence: New ideas from the Global South. African Minds,* Cape Town, p. 1–18.
43. See Polanyi (1962), p. 10.
44. Ibid., p. 15.
45. Op cit.
46. Cash, D. W., et al. (2003). "Knowledge systems for sustainable development." *Proceedings of the National Academy of Sciences of the United States of America* 100(14): p. 8086–8091.
47. Kallis, G. (2018) *Degrowth. Agenda Publishing,* Newcastle Upon Tyne, UK. p. 1.
48. Lane, R. (2019) The American anthropocene: Economic scarcity and growth during the great acceleration. *Geoforum* 99, 11–21. p. 12.
49. Walker, J. (2020) *More Heat than Life: The Tangled Roots of Ecology, Energy, and Economics*. Springer Books. p. 4.
50. Lane (2019) and Warde, P., Robin, L. and Sorlin, S. (2019) *The Environment*. John Hopkins, Baltimore.
51. Walker, J. (2007). *Economy of Nature: A Genealogy of the Concepts "Growth" and "Equilibrium" as Artefacts of Metaphorical Exchange between the Natural and the Social Sciences*. University of Technology, Sydney, Sydney. P. 19.
52. Coffey, B. (2016) Unpacking the politics of natural capital and economic metaphors in environmental policy discourse. *Environmental Politics* 25, p. 203–222.
53. Koopman, C. (2010). Revising Foucault: The history and critique of modernity. *Philosophy & Social Criticism* 36(5): p. 545–565, p. 556, discussing the work of Michel Foucault.
54. Koopman, C. (2010) Revising Foucault: The history and critique of modernity. *Philosophy & Social Criticism* 36, 545–565. p. 557.

55. Massumi, B. (2018) 99 *Theses on the Revaluation of Value: A postcapitalist manifesto*. University of Minnesota Press, Minneapolis.
56. Eaton, C., Stevens, M.L. (2020) Universities as peculiar organizations. *Sociology Compass* 14, p. 12768.
57. Barnett, C. (2018) The financialisation of higher education and the USS pension dispute. *Medium*. April 11 2018, https://medium.com/uss-briefs/the-financialisation-of-higher-education-and-the-uss-dispute-9231b9458699
58. See Eaton and Stevens (2020).

6

Learning and Teaching Matters

Advancing Sustainability

Our approach to the SDGs in higher education seeks to open up new ways of thinking about their reciprocal relationship as both an historical trajectory (that has over-emphasised development) and a newly defined critical movement focused on the need for greater sustainability, ecological integrity and social justice. As a 'politics of collaborative entanglement'[1] this involves focusing on how universities are intimately enmeshed and intertwined with the SDG agenda, and how this can be co-developed in more progressive, generative ways. This entails greater collective responsibility for developing a critically reflexive pedagogical praxis that is cognisant on and responsive to the precariousness of humanity's situation and the transformative change needed to support and sustain both people and planet. As David Orr describes:

> We must come to see ourselves as implicated in the world, not simply isolated, self-maximising individuals ... most education simply reinforces practices and pathologies that cannot and should not be sustained over the long term. ... This requires a new understanding of ourselves and our place in nature and time. This is *the* challenge of education.[2]

Unlike research, education (including higher education) is explicitly mentioned in the SDG agenda. In particular, it is the subject of Goal 4: Quality Education. Target 3 specifies: 'By 2030, ensure equal access for all women and men to affordable and quality technical, vocational and tertiary education, including university'.[3] Part of the driver for expanded access is the demand to ensure more people develop the vocational skills and training needed to succeed in their career, as SDG 8 on Decent Work and Economic Growth also aims for. But rather than just trying to provide labour for capitalism, SDG 4 also significantly includes a call to greatly expand teaching on sustainable development:

> By 2030, ensure that all learners acquire the knowledge and skills needed to promote sustainable development, including, among others, through Education for Sustainable Development and sustainable lifestyles, human rights, gender equality, promotion of a culture of peace and non-violence, global citizenship and appreciation of cultural diversity and of culture's contribution to sustainable development.[4]

It is significant and encouraging that the SDGs specifically call for the teaching of Education for Sustainable Development (ESD). A rich and evolving field,[5] ESD draws in part on the work of Brazilian activist Paulo Freire's landmark publication *Pedagogy of the Oppressed* which remains a seminal text in critical education for the environment. Freire's vision of critical pedagogy focuses not just on the 'what' of learning and teaching (L&T) content, but on 'how' education can be an accessible and transformative process for all who participate. His process of *conscientisation* emphasises the need to build critical awareness through reflection and action focused on 'learning as a critical process which depends upon uncovering real problems and actual needs'.[6]

Freire argues that 'neoliberal doctrine seeks to limit education to technological practice' and exacerbates an institutional context within which the 'opportunities for change become invisible, and our role in fostering change becomes absent'.[7] Critical pedagogy emphasises 'education for the greater good' and aims to cultivate education as 'a seedbed for new knowledge and culture leading to new selves, new societies, and a new humanity that is more humane'. Critical pedagogy can also help educators and students respond emotionally and intellectually to the proliferating array of contemporary problems and needs. As anthropologist Boone Shear describes:

> ...students—like many of us—feel and know deeply in their bodies that there is something terribly and fundamentally wrong with the reality that they are in, that it is losing coherence and meaning; and the narratives that offered a sense of individual and societal purpose are unable to provide psychic and somatic relief from a precarious world. ... More studies, more critique, more expert knowledge about why this is happening will not save us. I am interested in an approach that is more about ... ourselves as otherwise, as part of other emerging worlds that are not circumscribed or dominated by capitalist modernity, and that might help ... us learn to survive well, or at least survive.[8]

Among the 'emerging worlds' that critical pedagogical approaches try to cultivate, are communities of practice that embrace and support both the critically reflective practitioner and the practical scholar to better address local and global-scale sustainability challenges. This includes actively seeking and creating the opportunities to expand the available knowledge base, enhance democratic opportunity and debate; recognise different forms of local knowledge, history and politics, and co-develop the necessary skills and practices to enact the transformative vision of the SDGs.[9] It is also about celebrating what distinguishes the great capacity of education by leveraging the power of deep knowledge, academic networks and interdependence in order to create the possibilities for transformative change. This approach to education 'stands in stark contrast to current neoliberal trends that seek to privatize, standardize and script curriculum and pedagogy, and otherwise de-skill and disenfranchise teachers and students'.[10]

> Teaching and learning that understands itself as amplifying—not just understanding and critiquing—reality, can more intentionally untangle from the world as it currently exists and entangle itself into a collaborative politics of ontological possibility.[11]

In this chapter we turn our focus to the significance and importance of Learning and Teaching (L&T) in higher education *about, for* and *through* the SDGs. In the following sections we engage with the rich resources offered by earlier work on the education-sustainable development relationship, and with some of the many ways in which critical pedagogy is emerging and developing within what some describe as the 'posthuman turn' to higher education. Universities not only do practical applied educational 'work' of the sort many actors can do; they have a privileged capacity to identify and amplify neglected issues and voices, articulate lessons from the past, critique existing approaches, as well as anticipate possible futures. This includes the accelerating shift to online and virtual modes of engagement further propelled by the climate emergency and COVID-19. In this way universities can help both untangle problems and entangle like-minded people through critical and reflexive pedagogy.

Education *About, For* and *Through* the SDGs

A defining characteristic of the SDGs is that the issues and knowledges it encompasses are deeply entangled. Yet pioneers in early iterations of Environmental Education (EE), Education for Sustainability (EFS) and more recently Higher Education for Sustainable Development (HESD) highlight the tendency for educators to focus on just one of the pillars of the sustainable development triangle—social equity, economy *or* environment—rather than the critical interplay between all three. In the sustainability education field, the focus is for example, on the ecological footprint in cities and on campuses, issues of environmental justice and/or the impacts of neoliberalism on the environment. These are complex issues which require an interdisciplinary approach to sustainable development, yet there is a strong tendency in higher education pedagogy and practice to create silos for environmental education. Addressing the inter-relationships between all three of the key pillars in the 'sustainable development triangle' outlined in the 1987 Brundtland report, 'Our Common Future' is critically important.[12]

The elusiveness of the centre of the sustainable development triangle (see Fig. 6.1) has been discussed by Scott Campbell. Pointing to the absence of mechanisms steering education towards the centre, he calls for more explicit attention to the silences and tensions in the spaces in, and

6 Learning and Teaching Matters   173

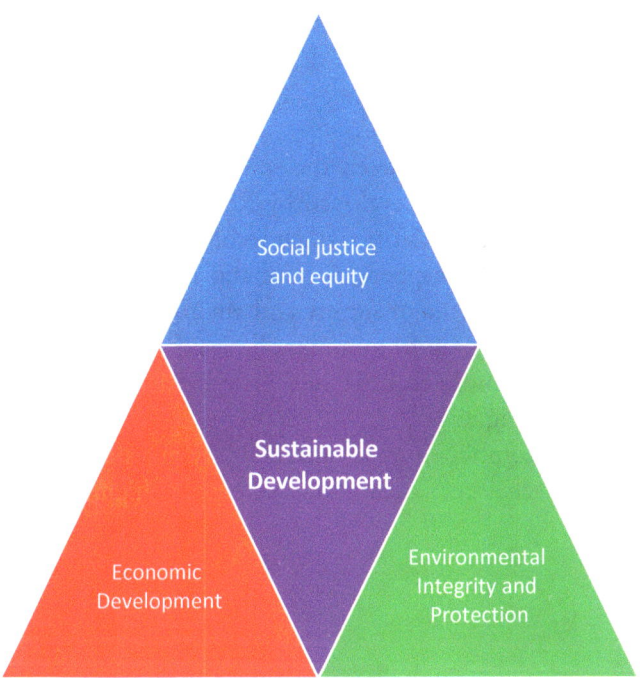

Fig. 6.1 The sustainable development triangle

between, social justice, economic growth and efficient and environmental protection. To address the triangle's inherent conflicts, he suggests, the focus must be on finding better ways to 'integrate social theory with environmental thinking, as well as combine techniques for community conflict resolution in order to confront economic and environmental injustice'.[13]

One of the strengths of the SDGs as a transformative pathway and pedagogical L&T agenda is making more visible the different dimensions of sustainable development across and between the seventeen goals. The increase in the number of goals does not suggest the tensions and contradictions between them have been resolved, as seen for example between *SDG13 Climate Action, SDG9 Industry, Innovation and Infrastructure* and *SDG11 Sustainable Cities* and Communities. Instead, the emphasis of the SDGs is on amplifying understanding around the complexity of sustainable development which can then be mobilised to support the achievement of more sustainable development—for example between *SDG7 Affordable and Clean Energy*, *SDG10 Reduced Inequalities* and *SDG17 Partnerships for the Goals.*

The interrelationships between the seventeen goals are illustrated in the 'wedding cake' model developed by Rockström and Sukhdev from the Stockholm Resilience Centre which reconceptualises the goals across the three domains of biosphere, society and economy. With economy represented as an outcome of society and environment, the model is an ambitious transition away from an emphasis on anthropocentric development towards a more realistic biosphere-led framing of sustainable development (see Fig. 6.2).[14]

The different combinations of sustainability goals are endless and so too are approaches to them. Some people highlight the goals'

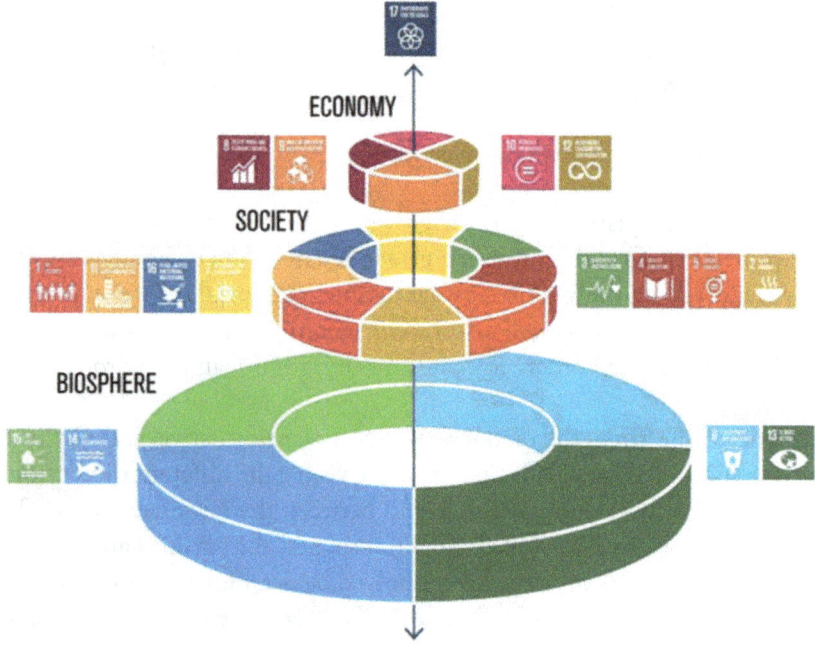

Fig. 6.2 Sustainable Development Goals 2030 (SDGs) wedding cake. (Source: Rockström and Sukhdev, Stockholm Resilience Centre, 2016)

interrelationships and explore cross-cutting themes, whilst others do a deep dive into a particular issue via just one of the goals. A diversity of critical pedagogy approaches to the SDGs is needed and represents in and of itself the sort of creative and inclusive approaches that are required. In contrast to an instrumentalist, institutional or technical agenda for universities, this involves addressing the SDGs through a focus on the 'interactions between humans and non-human nature—and how we relate to and respect the rights and dignity of these'.[15]

It also involves taking seriously the practical, physical and political implications of the SDGs for any university in which such L&T is conducted. This means helping ensure that students are 'experiencing concrete sustainable practices [...] in the daily campus operations' by collectively pushing and assisting the university to practice 'what it preaches in its classrooms'. In this way, universities can help grow what otherwise may be just a specialist topic into part of a broad 'value transition' within the institution, and to cultivate a 'civic sense' not only among students but among staff too in keeping with the ideal of the civic university which is discussed in Chap. 3.[16]

Above and beyond the field of Education for Sustainable Development (ESD) per se, critically engaged and reflexive understandings and practices around L&T and their relationship to the SDGs are evolving in higher education to address the need for meaningful real-world change. As educators this is an opportunity to attune to what is most important and to do what educators do best. This does not mean it is easy, but instead requires pausing to ask hard questions about what the world needs, rather than simply what the market wants. Core questions underpinning critical higher education *about, for* and *through* the SDGs include:

- What sort of education about sustainable development in higher education is needed? Who decides such questions?
- How do they justify such decisions?
- By what criteria and authority?
- With what pedagogical frameworks and practices?

Universities have the opportunity to harness existing and latent L&T potential to address and meaningfully engage with the SDGs, and foster the innovative cross-scale, cross-sectoral linkages, creativity and experimentation needed to successfully pursue a transformative sustainable development agenda. However, this involves identifying 'who is doing what' in terms of engaging with the SDGs to inform curriculum transformation, and fostering the strategic discussions needed in order to understand where strengths and silences lie, and to develop more informed, strategic and targeted interventions to better embed the SDGs in L&T in the future. Such steps are deceptively simple. To help navigate their complexity, the following three critical frames (Fig. 6.3) offer a useful heuristic for universities. This builds on and extends previous ESD frameworks that have focused on education 'about' (content-based sustainability literacy), 'for' (critical questioning of assumptions) and education 'as' (a shift of worldview) sustainability.[17] Each will be described in turn.

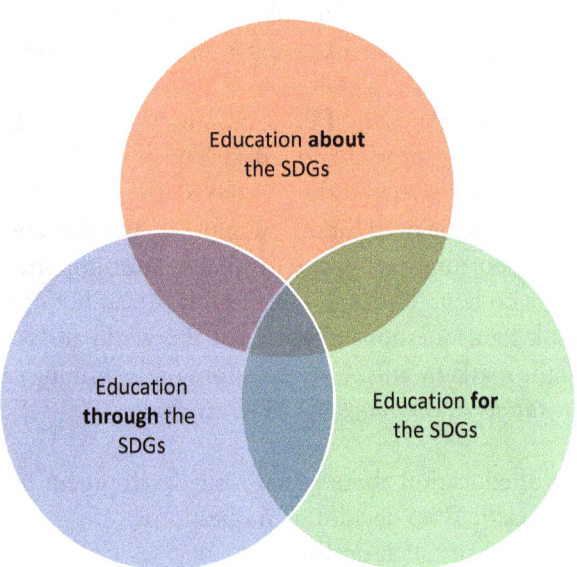

Fig. 6.3 Three critical frames for SDGs learning and teaching. (Source: Authors)

Higher Education *About* the SDGs

This first frame focuses on building knowledge and understanding of the SDGs framework itself. This is an explicit emphasis on the SDGs as a transformative agenda consisting of 17 goals and 300 associated indicators. The aim is the development of a shared language and platform for collectively addressing sustainable development. For example, the *Getting Started with the SDGs in Universities* report by the Sustainable Development Solutions Network (SDSN) Australia/Pacific calls, among other things, for educators to help students develop 'a basic understanding of the subject areas of each of the SDGs' and 'knowledge and understanding of the SDG framework purpose and uses'. Besides devoted units, education about the SDGs can be encouraged across the curriculum of a course of programme. It can also be combined with a critical praxis that actively explores the transformative potential of the SDGs as a United Nations (UN)-led framework and encourages students to work on ways to redress some of the notable silences in the SDGs such as Indigenous sovereignty and other critical development issues.

Higher Education *for* the SDGs

In this second frame a more holistic approach to the challenges and opportunities of *sustainable* development is taken. This frame underlines the need for critical pedagogy and praxis that addresses the big issues and improves the capacity of society to address issues of equity and justice and of the Earth to sustain life. Working across the boundaries of individual SDGs and beyond the SDGs as a stand-alone UN agenda, this approach explores the issue of sustainable development, associated 'developmentalities' and potential for transformational systems change. Pedagogically it brings into focus critical questions such as: who or what determines dominant ways of understanding and practicing sustainable development and prioritises objects for change? What are the contexts and pathways for transformative change, and how are they shifting?[18]

Higher Education *Through* the SDGs

The third frame recognises the benefits of undertaking a deep dive into particular goals as a way to address and leverage bigger societal issues. An example of this might be a strategic curriculum emphasis on gender and equity, quality education or urgent action on climate change. At the programme or course level, the pedagogical activities and assessment may emphasise sustainability within the context of different disciplines (e.g. business, architecture or science). Linking together several goals to engage with more intensively may also help to reveal tensions and potentialities that then speak out to the broader sustainable development agenda (e.g. Clean and Affordable Energy (SDG7), Industry, Innovation and Infrastructure (SDG9) and Responsible Consumption and Production (SDG12)).

All three frames are needed to embed the SDGs into the university L&T curriculum. As a stand-alone focus of the SDGs, SDG4 on Quality Education is crucial to the 2030 agenda's overall success. Utilising SDG4 as a pedagogical portal to the other sixteen goals may include, for example, a focus on SDG3 (Good Health and Wellbeing) within medical and health science courses, SDG6 (Clean Water and Sanitation) within engineering subjects, SDG11 (Sustainable Cities and Communities) within urban planning or sustainability studies, or SDG 16 (Peace, Justice and Strong Institutions) within law faculties. Other SDGs, such as SDG 10 (Reduced Inequalities) and SDG 13 (Climate Action), cut across different subject areas and can be designed to link together the discipline-specific L&T approaches within university faculties. In turn, an emphasis on the relations between different SDGs and differences between disciplinary perspectives could be combined with practical, collaborative action to progress an Education *for* Sustainable Development approach.

In pursuing Higher Education *about*, *for* and *through* Sustainable Development it is important to reflect on what is needed to help advance the SDG agenda. Among other things the following are recognised as important:

- students with the knowledge, skills and motivation to understand and address the history and complexity of the SDGs;
- creative, in-depth academic or vocational expertise to effectively implement SDG solutions;
- accessible, affordable and inclusive education for all;
- capacity building for students and professionals from developing countries;
- and the capacity to empower and mobilise people and their communities.[19]

While the SDG agenda uses 'community' in a conventional place-based way, empowerment and mobilisation for the SDGs is needed across many different types of communities, including professional, virtual and more-than-human ones.[20] Thinking of communities in this more expansive way opens up opportunities for innovative pedagogical action within L&T, of the sort we now discuss.

Pedagogical Innovation and the SDGs

Our approach to the SDGs as a transformative framework is that it does not only need to be innovative but underpinned by 'ethical innovation' that is Responsible, Attentive, Disruptive, Authentic and Regenerative (RADAR) (which we discussed in Chap. 4). A commitment to *ethical* innovation and not just any innovation for innovation's sake is particularly important in the face of the growing pressures on universities. Consistent with ethical innovation are efforts to cultivate opportunities for staff and students to actively learn *about, for* and *through* the SDGs, as we have outlined above. To foster such opportunities and collective engagement, institutional recognition and support of the positive benefit and worthiness of such effort is needed (e.g. by supporting, rewarding and developing a community of SDGs practice within a student-centred practice-based ethos).

Layers of ethical innovation are needed here, from institutional structures and processes around L&T (including how L&T relates to other parts of the university), to innovations in curriculum design, assessment and pedagogy. Each site for innovation needs to be partially shaped by how it can enable innovations elsewhere to help ensure it is as disruptive and regenerative as possible. In this way, innovation in L&T can help create the sort of emerging worlds Boone Shear discusses in the introductory section above. But this is not only about developing ethical innovation practices in universities or the higher education sector. It is about participating in and cultivating positive change in the multiple places and at the multiple scales that we are inescapably entangled with, including our domestic spaces, nation states and of course the Earth itself. As Dominic Orr asks rhetorically in the spirit of Paulo Freire's critical conscientisation after all: '*what good is innovative pedagogy if you don't have a decent planet to put it on?*'[21]

Selecting which approaches to use for advancing sustainability in higher education reflects is complex, reflecting the diverse questions, issues and problems that could be tackled. For this reason, the SDGs amplify the need for creative and eclectic approaches to L&T pedagogy. Sustainability educator John Fien's response to this dilemma is two-fold:

1. Consideration of both human consciousness and political action to answer moral and social questions about educational programmes which the dominant form cannot; and
2. Approaches that encourage critical and creative thinkers to take responsibility for their actions and participate in the social and political reconstructions required to deal with social/environmental issues within mutually interdependent and evolving social situations.[22]

Fien's emphasis on responsibility resonates with our call for ethical innovation. Besides creative and eclectic L&T initiatives, the SDG agenda requires *ethically* innovative L&T. This is a commitment to transformative educational change through deep learning and interdisciplinarity as opposed to isolated patches of innovation that academic educators Kathryn Hegarty and Sarah Holdsworth describe.[23]

The transformative agenda of the SDGs positions universities as catalysts for the development of new mentalities, capabilities and behaviours that underpin eco-social change. This is an emphasis on the critical importance of not just 'what is learned' but also 'how it is learned'.

Pedagogical innovation can be described as 'an intentional action that aims to introduce something original into a given context and pedagogical as it seeks to substantially improve student learning in a situation of interaction and inter-activity'.[24] Anne Walder suggests it can be pursued by:

1. Reflexively recognising the antecedents and need for change;
2. Designing, experimenting with and co-delivering an appropriate and beneficial pedagogical disruption and/or intervention;
3. Refining new structures, activities, approaches and initiatives through feedback with change agents in situ.

Other studies suggest that pedagogical innovation has six key characteristics: novelty, change, reflection, improvement, human relations and technology.[25] While the term improvement and related focus on quality points to overlaps with our call for ethical innovation, what counts as improvement is not simple or self-evident and this framework of pedagogical innovation can be extended by a more explicit emphasis on ethical innovation using our five RADAR principles in order to ground and orientate L&T endeavours (see Table 6.1 below). As with any sort of

Table 6.1 Ethical pedagogical innovation

Pedagogical innovation	Characteristics	Ethical innovation principles
Novelty	Counterculture, surprising, different, new	(R)Responsible?
Change	Slight-radical, incremental—transformative	(A)Attentive?
Reflection	Critical engagement, reflexivity, application	(D)Disruptive?
Improvement	Quality, indicators of success, feedback loops	(A)Authentic?
Human relations	Learning, opportunity, relationships, risk-taking	(R)Regenerative?
Technology	Use and access, links to pedagogy	
Precautionary	**Not all innovation is good, needed or useful**	

innovation, not all pedagogical innovation is necessarily good, useful or needed, as the proliferation of educational apps arguably demonstrates.

An example of how a structured ethical innovation approach can enhance existing pedagogical innovations around the SDGs is the use of pedagogies of 'discomfort'. Such pedagogies embrace conflict, different views and emotions to help students explore their entanglements in the problems underpinning the need for the SDG agenda, such as the climate emergency and systemic global inequality and poverty. They provide a pedagogically innovative approach to addressing the complexity of the SDGs 2030 agenda and action on it. Initiatives can include creative approaches to 'opening the classroom door' to what is in the room (emotions, resistance, desires, fears) through class discussion or various forms of social and artistic mediums, such as theatre, counter-mapping and improvisation.[26]

To be an effective, regenerative strategy, however, this needs to address the ethical innovation principles of RADAR to ensure that it is underpinned by care, training and empathy and avoids inflicting damage, hurt or harm. The ethical principle of attentiveness is important here in reminding us that what is appropriate can be highly specific to given contexts and groups. Acknowledging diversity in people's circumstances, perspectives and needs is one of the things the SDG agenda calls on us (albeit imperfectly) to do. More specifically, to provide inclusive, quality education that helps empower and mobilise people and their communities of the sort SDG 4 calls for, sensitivity to cultural, gender and social differences is paramount.

The development of innovative pedagogical capacity around the SDGs in higher education involves strategic and ethical choices about where, how and why to be innovative. Such choices can build on and leverage off existing institutional strengths in order to develop effective strategies and practices in L&T, identifying what is good and special about an institution, for example, and identifying what can be done better or differently to support sustainable development. Counteracting knee-jerk responses to market forces, this can strengthen commitment and help activate the university's public mission and civic orientation, and in doing so grow the space in higher education for L&T alternatives.

The SDGs focus attention on the need for a higher education system that supports sustainable development thinking and practices. As Ingrid Mula and colleagues argue, part of this is about prioritising 'the education of educators—building their understanding of sustainability and their ability to transform curriculum and wider learning opportunities'.[27] This is a process which as outlined above, demands ethics, creativity, inventiveness and initiative to deliver 'a better way of doing education'. It also requires both top-down (from senior-executive management) and bottom-up (from students and teachers) leadership to enable transformative change.[28] The challenge is to foster ethical pedagogical innovation without stifling 'bright spots' through a rigid, prescriptive and/or overly bureaucratic approach.

Applying the critical research impact work of geographer Ruth Machen is instructive here for re-thinking what ethically innovative L&T pedagogy might focus on, and how curriculum and assessment frameworks can support, 'rather than squeeze out', spaces for the SDGs as a transformative, reciprocal agenda. She outlines four routes to creating critical impact that can be applied to higher education L&T *about, for* and *through* the SDGs. These include,

1. *Challenging policy:* amending, changing or highlighting the implications of mainstream L&T/SDGs policy;
2. *Empowering resistances*: building L&T knowledge, networks and tools;
3. *Platforming voices*: listening, supporting, representing, mobilising marginalised voices; and
4. *Nurturing new critical publics*: inspiring critical skills and new forms of critical engagement.[29]

For critical L&T pedagogy and higher education *about, for* and t*hrough* the SDGs, making more visible the unhelpful political and discursive work performed through dominant pedagogies and mainstream curriculum is a central aim. Machen's work raises important questions 'about the relationship between knowledge translation and hegemonic power, and the way that this relationship shapes particular forms of neoliberal climate governance'.[30] The dominant development discourses and delivery modes driving higher education L&T are problematised in this approach by bringing to the need to address such knowledge practices and politics as part of meaningfully addressing the SDGs.

Machen cautions, however, that critical approaches can be co-opted, redirected, silenced and/or sabotaged. For one thing, they can be difficult to fit to those institutional structures and policies that have yet to be innovatively aligned with SDG L&T. For example, retro-fitting innovative, critical learning and teaching to existing assessment requirements can be challenging, while one of the side-effects of 'pedagogies of discomfort' (mentioned above) is that they are disliked by some students and rated poorly in evaluations. More broadly, as Machen notes:

> critical approaches impassioned by a desire for social change and seeking to challenge the status quo by unpacking the socio-historical contingency of meanings and exposing the reproduction of structural inequalities of power—often face a more challenging pathway to impact.[31]

Yet, as mentioned above, shifts in thinking and perspective can generate positive impacts that simply remain below the radar or outside of the timeframe of formal assessments, contributing to improved L&T practices and discourses and helping normalise less unsustainable ways of being.

Creating an Enabling SDGs L&T Environment

A critical L&T praxis *about, for* and *through* the SDGs requires more than changes in the classroom; it requires a critical praxis both inside and outside the formal curriculum and across different disciplines and global, regional and national geographies. It also requires institution-wide efforts to create the right enabling environment. Research in ESD and related fields indicates that a series of shifts are required: (1) from SDGs mapping to curriculum embedding; (2) from the delivery of expert-led content and teacher-focused pedagogies to collaborative problem-solving and student-focused pedagogies; and (3) from a focus on inputs to graduate outcomes. Each of these shifts will be outlined below.

From SDGs Mapping to Curriculum Embedding

An early step identified in the *Getting Started with the SDGs* report is curriculum mapping to outline the current 'state of play' with the uptake and integration of the SDGs within L&T programmes, courses and activities. Once such information is attained, attention can turn to ascertaining reasons for existing silences and gaps on the SDGs. As Raewyn Connell reminds us, 'every curriculum is a selection of the available knowledge'.[32] Importantly, even the notion of mapping the SDGs may be a barrier here, given that much teaching about and especially for them may not be detectable by standard word search approaches to mapping. There is tension between the hegemonic curriculum which is orderly, closed and full of answered questions, and the world in crisis which is ragged and increasingly full of unanswered questions that the SDGs—as an enormous live issue, not neat curriculum package—require we attend to.

Even if SDG content 'is' mappable, the challenge is to then expand and *embed* its presence in the curricula, keeping in mind the need for Education *about, for* and *through* the SDGs (see above). Key to infusing the SDGs as a transformative agenda through the curriculum are interdisciplinary approaches that call to the fore differences between and the assumptions and value of specific disciplines, explore the insights each offers about the SDG agenda, and in doing so work to combine such perspectives and approaches to a greater or lesser degree. If done effectively, interdisciplinary approaches can enable more open, outward-facing, cooperative, participatory, social and problem-based approaches to L&T to flourish, including on the SDGs.[33] Encouraging interdisciplinary research and L&T through various initiatives such as new centres or platforms is now a recognised mechanism for trying to support SDG engagement within universities.

Beyond the formal curriculum, there is a need to embed the SDG agenda within L&T practices and the higher education culture more broadly. Among other things, this requires transformation in some L&T delivery and modes—not in a narrowly prescriptive way, but in a way that embeds interdisciplinary and transdisciplinary approaches, develops communities of practice, and fosters cross-faculty and cross-institutional networking and collaboration opportunities for staff and students.[34]

Identified challenges to embedding the SDGs include the absence of a strategic framework around the SDGs and low levels of existing SDG and sustainability literacy and capabilities within among university staff.[35] Thus staff training and professional development is one of the institutional pillars needed to embed the SDG agenda in higher education.

From Teacher-Focused to Student-Focused Pedagogies

We come then to the second general shift required: from expert-led content and teacher-focused pedagogies to collaborative problem-solving and student-focused pedagogies. Problem-based learning is of clear relevance to the real-world issue of the SDG agenda. Like interdisciplinarity, which is often motivated by the need for disciplines to combine forces on applied problems,[36] it reflects pressure on universities to structure their knowledge generation and education around things of clear value to wider society (as discussed in Chap. 3). This includes future and existing students, as well as graduates. Real-world problem and engagement models are often associated with student-centred learning approaches, including experiential learning.[37] These are favoured in education on sustainable development and other complex topics as a means of generating transformative, classroom-based educational experiences.[38] Active, self-directed and collaborative learning that fosters reflection, self-awareness, empathy and shared experiences is also often used in these approaches, including reflection on which problems and whose problems count.[39]

The Futures Literacy Laboratory (FLL)[40] provides an example. It is a novel experience-based and process-based pedagogical model that focuses on encouraging students 'to use the future in particular ways' by making their anticipatory assumptions explicit. The FLL consists of three phases which include *Reveal* (from tacit to explicit and making assumptions about the future explicit); *Reframe* (identifying the key 'Aha' moments and insights); and *Rethink* (awareness of nuances, new questions, framings and lines of flight). The starting point for futures literacy is perceiving and understanding anticipatory assumptions to interrupt the routine action of 'using-the-future'. Through action learning the aim is to reveal not only the determinants of the futures they imagine, but also the attributes of the anticipatory systems and knowledge creation processes that they use when thinking about the future. This is a structured process that involves a number of key steps which could usefully be adapted for critical engagement with the SDGs through the development of FLL-SDGs.

1. Participants experience and become explicitly conscious of how the future plays a central role in what they perceive and pay attention to in the present.
2. By changing the ways that they 'use-the-future' participants start to realise they can anticipate in different ways and thereby imagine different futures.
3. By putting together the first and second insights participants begin to understand that imagining different futures changes what they could see and do in the present.
4. By imagining different futures participants become more aware of their own capacity to invent the underlying anticipatory assumptions that shape their descriptions of the 'later-than-now'.
5. By starting to acquire futures literacy they become better at rooting their anticipatory assumptions in their own history and specific socioeconomic-cultural context. Participants begin to reassess their perceptions of the present, depictions of the past and aspirations for the future.
6. Through engagement in the knowledge co-creation processes participants begin to acquire the capacity to design this kind of collective intelligence process that enables them to choose why and how to anticipate, contributing to the acquisition of the skills that make up futures literacy.[41]

Another quite different example of innovative student-centred praxis is through the eco-pedagogy that environmental/outdoor educators such as Gen Blades write about and practice. Her work focuses on the sensory and affective dimensions of 'walking in nature' as a means by which to critically engage with the patriarchal, instrumental and commodified nature and orthodoxies of mainstream education. The twofold purpose of this approach is to: (1) *decentre* the role of humans amidst the complex entanglements of shifting human-nature relationships; and (2) to purposefully privilege the role of embodied experiences in L&T as a means by which to *emplace* the more-than-human in alternative curriculum modes and models.[42]

From Inputs to Graduate Outcomes

The third shift associated with an SDG focus in L&T is a change in emphasis from inputs to graduate outcomes. Graduate outcomes are given weight because they are taken to represent the sum total of an institutions' influence upon a student. These influences include not just classes but informal learning experiences such as study tours, hackathons, conferences, youth training and community leadership programmes that can all usefully be infused with an SDG focus.[43] Thinking about graduate outcomes re-focuses attention on the aim of SDG education, which can be described as:

> the co-creation of learning environments and opportunities that support learning on the SDGs … and … structure courses around real-world collaborative projects for change, in which the students have the opportunities to act and reflect iteratively, and to develop adaptive capacity while working towards a purpose.[44]

Terms such as 'act', 'reflect' and 'adaptive' gesture to the character of the graduate outcomes considered valuable for the SDG agenda. SDG Target 4.7 explicitly encourages the development of key cross-cutting skills or competences in students, where the latter refers to an individual's set of demonstrable skills and characteristics that are critical for sustainable futures.[45] Key competences considered helpful in understanding and addressing the SDGs include skills in systems thinking, values thinking, action learning, interpersonal skills, strategic management and integrated problem-solving[46] (see Table 6.2 below).

A competence approach encompasses both specific technical or academic competences and 'transversal' competences or 'soft skills'. In terms of the SDGs, soft skills are recognised as essential for the implementation, communication and integration and mobilisation of the SDGs through critical citizenship and for helping enable the active, reflective

Table 6.2 SDG Skills and Competences

Competence	Definition
Systems thinking	Recognition and understanding of relationships and ability to analyse complex systems at different scales, and adaptability to uncertainty.
Anticipatory	Capacity to understand and evaluate multiple futures; ability to create one's own vision, apply the precautionary principle and deal with risk and emerging challenges.
Normative	Capacity to understand underlying norms and values; ability to navigate sustainability targets, goals, principles and necessary trade-offs.
Strategic	Ability to develop and implement collective actions that further local-level sustainability goals/visions.
Collaboration	Ability to learn from others and be empathetic and sensitive to different needs, perspectives and approaches; ability to facilitate collaboration and manage and resolve conflict.
Critical thinking	Ability to question practices, views and norms while being reflective on one's own values, actions and perspectives.
Self-awareness	Ability to reflect on one's own role at both local and global levels, self-motivate towards action(s) and manage feelings and desires.
Integrated problem-solving	Ability to use different approaches to problem-solving to respond to complex sustainability challenges and establish inclusive, practical and equitable solutions by integrating all competences above.
Digital	Ability to use digital software, programmes and devices to carry out sustainability activities and communicate SDG-related content to a high standard (Vialta et al. 2018).

Adapted from Rieckmann et al. 2017; SDSN 2017

and committed participation in L&T activities that helps cultivate it. Characterised by these and other competences, ideal graduates are insightful and self-aware, with the ability to imagine and help create different, more positive futures.[47]

Student-led and community-centred education of the sort discussed above is considered crucial to developing these competences. As Rieckmann and colleagues describe, 'competencies cannot be taught, but have to be developed by the learners themselves. They are acquired during action, on the basis of experience and reflection.'[48] Active student involvement can involve helping design the course content or engaging in practical SDG-related projects. The latter may not only build students' skills and knowledge but also enhance their capability and confidence to engage with SDG issues beyond the classroom.

This space of 'beyond the classroom' could be an urban garden or smart city on the other side of the world, or it could be the university they are part of. Skilling up students to engage critically with academic institutions and use their voice and efforts to help embed the SDGs at all levels and in all corners within them, could be one of the most important outcomes of SDG-focused education. It could also help generate valuable positive feedback as students prove themselves to be not only 'products' of SDG education, but also sources of institutional capability on the SDGs and effective 'external' pressure to further embed the goals in societal practices more broadly.

Embedding the SDGs in universities requires a plethora of diverse and loosely coordinated approaches and groups. From high-level strategic commitment to the SDGs or socialising students in a holistic SDG ethos, to the delegation of responsibility for SDG engagement onto individuals or the funding of small groups to undertake voluntary initiatives outside the bounds of formal L&T,[49] there is virtually no limit to what can be done to embed the SDGs into educational experiences in universities. In a given institutional context, the SDGs may even be most loudly associated with radical community-driven initiatives and manifestos for change, such as enacting the 'free university' model or other alternative higher education pathways in the spirit of *Buen vivir*.

Although there is virtually no limit to what can be done to bring the SDGs into L&T *in theory*, in practice there are of course a range of barriers, some of which have been detailed above. Crucially, the existence of such barriers can illuminate the actual need for SDG action *within* higher education institutions themselves. For example, a lack of time, resources and security among staff and students is a major obstacle to them being able to explore and pursue the SDGs alongside their other responsibilities, commitments and activities. It is *also* a reflection of deficiencies around decent work (SDG 8), economic equality (SDG 9) and good governance (SDG 17) within their institutions, underlining why their institutions should engage with the SDG agenda and the potentially far-reaching implications of them doing so.

We turn now to another angle on the SDGs in L&T—the increasing emphasis on online and virtual delivery and global engagement—that further pushes against a reduction of the SDGs to mere study area by again emphasising the context of education and how close SDG issues always are.

Online and Virtual SDGs L&T: Challenges and Opportunities

The SDGs require both local-scale and global collaboration if they are to be achieved by 2030. Technology is playing, and will continue to play, a central role in delivering the 2030 Agenda, particularly through the rise in digital communications, online learning and virtual collaboration as a result of the COVID-19 global health pandemic. Alongside the increasing role of online pedagogy and education focused on the SDGs in higher education, is growing recognition of the need for digital democracy and a response as part of the SDGs to the digital divide. That is, once again the sort of inequalities that the SDG agenda points to are inescapably part of the context of higher education, not because they are a distant issue to engage with, but because it is inscribed in its uneven geographies and relations.

Although seemingly ubiquitous—and thus invisible to most of us—digital technologies and infrastructure are far from evenly distributed across the world, spatially or socially.[50] Geographer Matthew Gandy argues that the world is increasingly characterised by 'global citadels of [digital/virtual] connectivity encased within a wider landscape of [material/urban] neglect and social polarization'.[51] While universities are frequently part of such citadels (especially those that pride themselves on being 'world class'), many of their current and prospective students are not. Nor their casualised staff.

Thus, efforts to address the SDGs through online means, such as a digital SDG-based L&T platform that seeks to teach students about issues such as energy justice, cannot dismiss the wider issues of inequities in access to affordable digital (and electricity) resources. An institution making an online program or course/subject on the SDGs *available* does not necessarily imply everyone will be able to access and benefit from it, with some for instance lacking the internet bandwidth and others lacking a safe space or time to study.

A range of factors can hinder access to online SDGs education including the language it is taught in, existing education background or geographical location.[52] If digital education allows some groups to accelerate ahead, while others are left behind, the proliferation of online education may increase social inequalities, even if the education being delivered is about the SDGs. At the same time, there are limitations and barriers to how effective online L&T can be for achieving SDG-related outcomes. While digital education can increase the efficacy and efficiency of pedagogical engagement and delivery relative to traditional face-to-face learning, these outcomes are far from guaranteed.[53]

Adding to uncertainty about the quality of online L&T is the tendency for many courses to still be designed in the conventional manner, both in terms of being designed by default for face-to-face and in terms of taking a top-down teaching approach. Neither translates well into the online environment. If this is how SDG knowledge, skills, competences and attributes are being developed in students, such efforts may raise students' awareness of the SDGs but not necessarily equip them with the tools for achieving the outcomes called for under SDG 4 or any other goals.

The rise of *Massive Open Online Courses* (MOOCs), for example, is a double-edged sword. They represent a step towards democratising education by making content available (often for free) to more people from more countries and engaging a larger proportion of the general public on a global scale on issues like sustainable development. Anyone with a computer (and electricity, internet, data, digital literacy, skills and confidence in the online world)[54] can access a MOOC. They are low cost and can be offered to students for free. MOOCs do not require students to have prior qualifications or experience so support lifelong learning and provide educational opportunities that foster intercultural perspectives and dialogue between different peoples, places and cultures.[55]

MOOCs are therefore available and accessible to a wider range of prospective students than normal university education. Experience to date indicates they can enable and encourage the formation of large global and diverse learning communities. Yet MOOCs do not provide formally recognised qualifications or accreditation, which can reduce their value to students needing external recognition of their skills and knowledge. Difficulties also arise around quality assurance of MOOCs and inherent language and cultural biases, depending in part on the teacher delivering the content and the demographics of the students participating.

But there are a lot of positives, including ones neglected in typical discourse about the online turn. Beyond just MOOCs, new spaces are being constantly created through the internet, generally making education more available and accessible. Online lectures and content provide flexibility for students with competing obligations outside of study like work, family and other commitments, while e-learning can increase motivation to learn, learning efficiency and learning success. Accessibility and flexibility in learning modes can also enable greater interdisciplinary and transdisciplinary approaches to L&T and boost lifelong learning—both of which are critical for SDG transformation.[56] The turn to online education and SDGs can also be mutually beneficial when they concurrently encourage exploratory, democratic, student-led approaches to L&T that engage with the dynamic world and students' lives.

An important development is rising awareness that online learning requires innovative L&T approaches besides the mere use of digital technologies. Approaching online learning as an afterthought or a nice 'add-on' to face-to-face education offerings is gradually giving way to approaches that centre and celebrate online learning as a worthwhile and meaningful stand-alone form of education. This shift to a stronger, stand-alone focus on digital education could help escape dominant, didactic approaches to SDGs L&T if accompanied by a parallel shift towards centring on the SDGs in and of themselves and fostering the ethical innovation outlined in the RADAR framework (see Chap. 4 and above).

A bold, innovative approach to online education is advocated by Sian Bayne and colleagues at the Centre for Research in Digital Education at the University of Edinburgh. Their *Manifesto for Teaching Online* features a series of provocative statements designed to counter 'both the "impoverished" vision of education being advanced and higher education's traditional view of online students and teachers as second-class citizens'.[57] This includes five thematic sections focused on: (1) examining place and identity; (2) politics and instrumentality; (3) the primacy of text and the ethics of remixing; (4) the way algorithms and analytics and educational intent work; and (5) how surveillance culture can be resisted. Their intention is to build a critical platform for those teaching in online environments to challenge the instrumentalism of current technology and digital pedagogical approaches. Several of their key provocations designed to both challenge and engage include:

- Distance is a positive principle, not a deficit
- Place is different, not less important online
- Many modes matter in representing academic knowledge
- Distance is temporal, affective, political, not merely spatial
- Online teaching need not be complicit with the instrumentalisation of education[58]

Digital learning innovations are also providing broader opportunities to meet the SDGs as a holistic agenda. They can increase students' understanding of the interrelation and interdependence between different SDG goals and help students build relevant skills and competences. They can also be used to gain a better idea of the diversity and complexity of practices around the SDGs through online participation by different groups of people with different types and levels of SDGs literacy, including across different parts of a university. In this way they can provide key tools and platforms for building SDGs-relevant knowledge, competences and attributes across broad segments of society, particularly during these uncertain times of global crisis.

In sum, critical engagement with online or digital learning (e-learning) can support the wider integration of SDGs into university L&T and its boundary crossing in space and time.[59] At the same time, it is important to remember that digital technology also presents a number of issues and limitations. Besides those mentioned so far, these include the health and wellbeing impacts of digital overload and social isolation on students and staff, and the greenhouse gas emissions and other externalities and resource scarcities exacerbated by inescapable material digital activities.[60]

Learning and Teaching Futures

As we have discussed and as this book exemplifies, momentum is rapidly growing around *Transforming Our World: The 2030 Agenda for Sustainable Development*, otherwise known as the SDGs. Laid out by the United Nations with the endorsement of all 193 of the nations it represents, this vision calls for concerted, integrated action on environmental sustainability, social justice and prosperity. It makes clear that society is at a key decision point on multiple intersecting challenges, including climate change and that how we collectively respond has far-reaching ramifications. Universities are recognised as having a leading role as agents of societal change, but they are themselves conflicted and implicated in the development legacies and innovation trajectories that have resulted in a 'world of wounds'.

In response, the 'posthuman turn' in education for sustainable development seeks to re-imagine the pedagogy, practice and research work of universities as critical assemblage. Central to this is how we can better engage with the situated knowledge, practices and ideas located at the nexus of bio (life), geo (earth) and techno (technology). Along with new modalities of education, posthuman education requires addressing the wider societal context within which higher education takes place. By overcoming some of the limits of the enlightenment tradition, posthumanism tries to open up new types of ethical engagements for critical pedagogy in education for sustainable development.

For Carol Taylor and colleagues, posthumanism represents a shift away from the enlightenment tradition that has focused Western education on 'the humanist cul-de-sac of individualism, binarism and colonialism' and the human-nature divide.[61] By challenging eurocentrism, masculinism and anthropocentrism, posthumanism encourages us to move forward into the complexities and paradoxes of our times. By re-framing the human endeavour towards realignment with both the vitality of matter and the more-than-human, posthumanism offers new critical, experimental and creative ways to do higher education differently. As Rosi Braidotti describes, posthumanism helps us to rethink, the historical moment when the Human has become a geological force capable of affecting all life on this planet and by extension help us to rethink the basic tenets of our interaction with both human and non-human agents on a planetary scale.[62]

Post-digital educator Sian Bayne (mentioned above) points to problems in conceiving of university education as the process for 'becoming fully human' if we do not take into consideration what this means or entails. Posthumanism she argues is a framework for thinking beyond the innate potential of students to the interdisciplinary tensions, synergies and systems that work together as agencements—'arrangements endowed with the capacity of acting in different ways depending on their configuration'.[63]

There is a need to critically reflect on the potential for a 'transformational' higher education committed to advancing the SDGs. Embedded with a critical praxis and building on the work of Education for Sustainable Development (ESD), transformational L&T approaches on, about and for the SDGs should not be prescriptive in form or content. Nevertheless, they are likely to encompass student-driven, interdisciplinary and boundary-crossing pedagogies and participatory approaches to knowledge, co-creation, generation and acquisition. In contrast to most purportedly apolitical approaches to education, transformational higher education for the SDGs is a critical, ethical agenda focused on transformative change-making in multiple arenas to help generate a more sustainable future.

Critical L&T pedagogy and praxis focused on the SDGs does not lend itself to a neat blueprint for curriculum design and assessment. It requires instead an approach that John Law refers to as method assemblage: 'the process of crafting, bundling together, gathering and enacting presence, absence and otherness'.[64] By this he means bringing together whatever is needed here in the present, while drawing attention to what is absent, repressed or hidden. A L&T-focused and SDGs-oriented method assemblage will arguably require three exploratory stages.

First, a transformational *mapping* of the possibilities for SDGs translation and creation within higher education. Mapping various bodies of work across SDGs themes and practices helps to address knowledge L&T

gaps in specific disciplines and research areas. Second, this provides a basis for undertaking a *diagrammatic* of the relational forces and tensions that are in play, as well as the possibilities for effective emergences and possibilities. Diagramming makes visible the interconnections and relations between agents (human and non-human) and helps to anticipate potential conflicts. Finally, directing effort and emphasis towards *sketching* specific examples of new L&T assemblages and potentialities that already exist or might be able to emerge as an iterative and relational process will provide creative impetus and direction.[65] Sketching (or imaginative thinking) allows for new lines of flight to address the SDGs. As Jean Hillier argues 'an analytic cartography inspired by Deleuze and Guattari can help to understand the micropolitics of power in connection with broader political, social and environmental structures and conditions of possibility'.[66]

While there are many examples of universities actively or incidentally seeking to integrate the SDGs into different areas of curricula and L&T, there is a need to question how 'useful' these initiatives are for building the competences, qualities and attributes that creates the progressive change needed to address the local and planetary scale issues we currently face. Many existing L&T initiatives focused on SDGs are top-down approaches that do not yet fulfil the need for student-led and transformative approaches of the sort the SDGs call for. There may be uptake around the SDGs, but often as a new product or discrete topic area, not integrated across L&T as the new *modus operandi*. As such, while these 'bright spot' initiatives might provide an introduction to students wishing to engage with the SDGs, they are unlikely to be fundamentally changing the way that staff or students are learning in practice.

Before designing, implementing or embedding L&T SDGs initiatives, universities must consider how they want staff and students—indeed the broader community—to be affected by their educational experience. Transformational approaches are likely to be future-focused and geared towards producing twenty-first-century competencies that include critical thinking and digital literacy skills, as well as systems thinking, values thinking, action learning, interpersonal skills, strategic management and integrated reflective practices and problem-solving. The emphasis following the critical pedagogy of Paulo Freire is not just on 'what' is the content, but 'how' L&T in higher education is undertaken, in what ways and for whom. We would extend this by also asking 'why' universities should embed the SDGs in L&T curriculum within the context of a rapidly heating and increasingly inequitable planet.

The SDGs as a critical, transformative agenda aims to 'move us in the direction of being healthier, safer, more productive individuals, and in a manner that protects our resources and planet for future generations'.[67] More than a list of problems or just an idea, it is an influential, galvanising driver of change across sectors. Universities have the opportunity and capacity to embrace a deep commitment to the SDGs, combined with a bold, ethical L&T innovation culture. This represents the *scaling up, out and deep* of the SDGs agenda from a niche concept into the everyday practices and ethos of higher education that underpins transformational sectoral change. Universities have the capacity and public mandate to lead on the SDGs to 2030 and beyond—the question is, will they?

Notes

1. Shear, Boone W (2019) "Toward an Ontological Politics of Collaborative Entanglement: *Teaching and Learning as Methods Assemblage.*" *Collaborative Anthropologies* 12(1), p. 50–75.
2. Orr, D (2001) Forward, in Sterling, S, *Sustainable Education: Revisioning learning and change*, Schumacher Briefing 6, Foxhole, Green Books for the Schumacher Society.
3. See https://www.undp.org/content/undp/en/home/sustainable-development-goals/goal-4-quality-education.html#targets
4. See SDG 4.7 Education for Sustainable Development and Global Citizenship, accessed online at https://www.coe.int/en/web/education/4.7-education-for-sustainable-development-and-global-citizenship
5. See Bart, M, Michelsen, G, Rieckmann, M and Thomas, I (2016) *Routledge Handbook of Higher Education for Sustainable Development*, New York, Routledge.
6. See discussion of Paulo Freire key concepts—https://www.freire.org/paulo-freire/concepts-used-by-paulo-freire/
7. Freire, P (1992) *Pedagogy of the Oppressed*, London, Penguin, p. 4.

8. Shear, Boone W. (2019) "Toward an Ontological Politics of Collaborative Entanglement: *Teaching and Learning as Methods Assemblage.*" *Collaborative Anthropologies* 12(1), p. 56.
9. Adapted from Healey, P. (2010) Re-thinking the relations between planning, state and market in unstable times, in Paolo, P and Vitor, O (Eds.) *Planning in times of uncertainty,* FEUP Edicoes, Porto, p. 15.
10. Macrine, S (2020) *Critical pedagogy in uncertain times: Hope and possibilities,* London, Palgrave Macmillan.
11. Shear, Boone W. "Toward an Ontological Politics of Collaborative Entanglement: *Teaching and Learning as Methods Assemblage.*" *Collaborative Anthropologies* 12(1), p. 75.
12. Brundtland Commission (1987) *Our Common Future, Oxford, Oxford University Press*
13. Campbell, S (1996) Green Cities, Growing Cities, Just Cities? Urban planning and the contradictions of sustainable development, *Journal of the American Planning Association*; Summer; 62 (3), p. 296–312.
14. See Rockstrom, J and Sukhdev, P (2016) https://www.stockholmresilience.org/research/research-news/2016-06-14-how-food-connects-all-the-sdgs.html
15. Fien, J (2002), Advancing sustainability in higher education: issues and opportunities for research, Higher education policy, 15, p. 143–152.
16. Sonetti, G, Brown, M and Naboni, E (2018) About the Triggering of UN Sustainable Development Goals and Regenerative Sustainability in Higher Education, *Sustainability,* 11(254), p. 1–17.
17. For a discussion of this see Lucas, A (1979) *Environment and Environmental Education: Conceptual issues and curriculum implications,* Melbourne, International Press and Publications. Sterling, S (2009) Sustainable Education, in Gray, D, Colucci-Gray, L, and Camino, E (eds.) *Science, Society and Sustainability: Education and empowerment for an uncertain world,* London, Routledge, p. 105–18.
18. Pelling, M, O'brien, K and Matyas, D (2015) Adaptation and Transformation, *Climate Change,* 133(1) p. 113–127.
19. SDSN Australia/Pacific (2017): Getting started with the SDGs in universities: A guide for universities, higher education institutions, and the academic sector. Australia, New Zealand and Pacific Edition. Sustainable Development Solutions Network—Australia/Pacific, Melbourne p. 2.
20. Rickards et al. (2019) Community resilience.

21. Orr, D. (1992). Ecological literacy: Education and the transition to a postmodern world. Albany: State University of New York Press.
22. Fien, J (2002) Advancing sustainability in higher education: issues and opportunities for research, in *Higher Education Policy* 15(2), p. 143–152. See also Robottom, I. and Hart, P. (1993). *Research in environmental education: Engaging the debate.* Geelong, Deakin University.
23. Hegarty, K and Holdsworth, S (2016) Towards a scholarship of curriculum change: From isolated innovation to transformation, in Barth, M, Michelsen, G, Rieckmann, M and Thomas, I (eds.) *Routledge handbook of Higher Education for Sustainable Development,* New York, Routledge.
24. Bechard, J (2000). *Learning to teach at higher education: the example of educational innovators.* OIPG Research Paper n° 2000-001, September, 6
25. Walder, A (2014) The Concept of Pedagogical Innovation in Higher Education. *Education Journal.* 3 (3), p. 195–202.
26. See Zembylas, M (2015) 'Pedagogy of discomfort' and its ethical implications: the tensions of ethical violence in social justice education, *Ethics and Education,* 10(2), p. 163–174; Kester, K, Zembylas, M, Sweeney, L, Lee, K, Kwon, S and Kwon, J (2019): Reflections on decolonizing peace education in Korea: a critique and some decolonial pedagogic strategies, *Teaching in Higher Education,* access online at https://d1wqtxts1xzle7.cloudfront.net/60202610/Reflections_on_decolonizing_peace_education_in_Korea_a_critique_and_some_decolonial_pedagogic_strategies.pdf
27. Mula, I, Tulbury, D, Ryan, A, Mader, M, Douha, J, Mader, C, Benayas, J, Dloughy, J, Alba, D (2017) Catalysing change in higher education for sustainable development: A review of professional development initiatives for university educators, *International Journal of Sustainability in Higher Education,* 18(5), p. 798–820.
28. Filho, L W, Shiel C, Paço A, Mifsud M, Ávila LV, Brandli LL, Molthan-Hill P, Pace P, Azeiteiro UM, Vargas VR, Caeiro S (2019) Sustainable Development Goals and sustainability teaching at universities: Falling behind or getting ahead of the pack? *Journal of Cleaner Production,* accessed online at https://repositorioaberto.uab.pt/bitstream/10400.2/8700/1/LEal2019.pdf
29. Ibid.
30. Machen, R (2018) Towards a critical politics of translation: (Re)-producing hegemonic climate governance, *Environment and Planning E—Nature and Space,* 1(4), p. 494–515.

31. Machen, R (2018) *Impact from critical research: what might it look like and what support is required?* LSE Blog, 4th September, accessed on https://blogs.lse.ac.uk/impactofsocialsciences/2018/09/04/impact-from-critical-research-what-might-it-look-like-and-what-support-is-required/
32. Connell, R (2019) *The Good University: What universities actually do and why it's time for radical change*, Melbourne, University of Monash Press.
33. Agbedahin, A.V. (2019), "Sustainable development, Education for Sustainable Development, and the 2030 Agenda for Sustainable Development: Emergence, efficacy, eminence, and future", *Sustainable Development*, Vol. 27 No. 4, pp. 669–80.
34. Mawonde, A. and Togo, M. (2019), "Implementation of SDGs at the University of South Africa", *International Journal of Sustainability in Higher Education*, Vol. 20 No. 5, pp. 932–50.
35. Filho, W (Ed) (2018) *Implementing Sustainability in the Curriculum of Universities: Approaches, methods and projects,* Springer Nature.
36. Barry, A., G. Born and G. Weszkalnys (2008). Logics of interdisciplinarity. *Economy and Society* 37: p. 20–49.
37. Brugmann, R., Cote, N., Postma, N., Shaw, E. A., Pal, D. and Robinson, J. B. (2019), "Expanding Student Engagement in Sustainability: Using SDG- and CEL-Focused Inventories to Transform Curriculum at the University of Toronto", *Sustainability*, Vol. 11 No. 2.
38. See Buil-Fabrega, M., Casanovas, M. M., Ruiz-Munzon, N. and Leal, W. (2019), "Flipped Classroom as an Active Learning Methodology in Sustainable Development Curricula", *Sustainability*, Vol. 11 No. 17; Melles, G. and Paixao-Barradas, S. (2019), "Sustainable Design Literacy: Developing and Piloting Sulitest Design Module", Chakrabarti, A. (Ed.), *Research into Design for a Connected World*, Springer Singapore, Singapore, pp. 539–49.
39. Barth, M. and Burandt, S. (2013), "Adding the "e-" to Learning for Sustainable Development: Challenges and innovation", *Sustainability*, Vol. 5, pp. 2609–22.
40. Miller, R., (2018). *Transforming the future: anticipation in the 21st century.* London, Taylor & Francis.
41. Ibid., p. 42.
42. Blades, G (2019) *Walking Practices with/in Nature(s) as Ecopedagogy in Outdoor Environmental Education: An Autophenomenographic Study,*

unpublished PhD thesis, School of Education, College of Arts, Social Sciences and Commerce, La Trobe University, Victoria, Australia.
43. SDSN Australia/Pacific (2017): Getting started with the SDGs in universities: A guide for universities, higher education institutions, and the academic sector. Australia, New Zealand and Pacific Edition. Sustainable Development Solutions Network—Australia/Pacific, Melbourne
44. Ibid., p. 13.
45. Demssie, Y. N., Wesselink, R., Biemans, H. J. A. and Mulder, M. (2019), "Think outside the European box: Identifying sustainability competencies for a base of the pyramid context", *Journal of Cleaner Production*, Vol. 221, pp. 828–38.
46. Moon, C. J., Walmsley, A. and Apostolopoulos, N. 2018, "Governance implications of the UN higher education sustainability initiative", *Corporate Governance—the International Journal of Business in Society*, Vol. 18 No. 4, pp. 624–34.
47. See Boluk, K. A., Cavaliere, C. T., and Duffy, L. N. (2019), "A pedagogical framework for the development of the critical tourism citizen", *Journal of Sustainable Tourism*, Vol. 27 No. 7, pp. 865–881, Straková, Z. and Cimermanová, I. (2018), "Critical Thinking Development-A Necessary Step in Higher Education Transformation towards Sustainability", *Sustainability*, Vol. 10 No. 10. Zamora-Polo, F. and Sanchez-Martin, J. (2019), "Teaching for a Better World. Sustainability and Sustainable Development Goals in the Construction of a Change-Maker University", *Sustainability*, Vol. 11 No. 15.
48. Riechmann, M., Mindt, L. and Gardiner, S. (2017), *Education for Sustainable Development Goal: Learning Objectives*, UNESCO, Paris, p. 10.
49. Zamora-Polo, F., Sanchez-Martin, J., Corrales-Serrano, M. and Espejo-Antunez, L. (2019), "What Do University Students Know about Sustainable Development Goals? A Realistic Approach to the Reception of this UN Program Amongst the Youth Population", *Sustainability*, Vol. 11 No. 13.
50. Graham, S. and S. Marvin (2001). *Splintering Urbanism: networked infrastructures, technological mobilities and the urban condition.* London, Routledge.
51. Gandy, M (2005) Cyborg Urbanization: Complexity and Monstrosity in the Contemporary City. *International Journal of Urban and Regional Research* 29.1 (2005): 26–49.

52. See Ahel, O. and Lingenau, K. (2020), "Opportunities and Challenges of Digitalization to Improve Access to Education for Sustainable Development in Higher Education", Leal Filho, W., Salvia, A.L., Pretorius, R.W., Brandli, LL, Manolas, E., Alves, F., Azeiteiro, U., Rogers, J., Shiel, C. and Do Paco, A. (Eds), *Universities as Living Labs for Sustainable Development: Supporting the Implementation of the Sustainable Development Goals*, Springer International Publishing, Cham, pp. 341–56; Bell, S., Douce, C., Caeiro, S., Teixeira, A., Martín-Aranda, R. and Otto, D. (2017), "Sustainability and distance learning: a diverse European experience?", *Open Learning: The Journal of Open, Distance and e-Learning*, Vol. 32 No. 2, pp. 95–102; Gallagher, S. (2018), "Development Education on a Massive Scale: Evaluations and Reflections on a Massive Open Online Course on Sustainable Development", *Policy & Practice: A Development Education Review*, Vol. 26, Spring, pp. 122–140.
53. Barth, M. and Burandt, S. (2013), "Adding the "e-" to Learning for Sustainable Development: Challenges and Innovation", *Sustainability*, Vol. 5, pp. 2609–22.
54. Gallagher, S. (2018), "Development Education on a Massive Scale: Evaluations and Reflections on a Massive Open Online Course on Sustainable Development", *Policy & Practice: A Development Education Review*, Vol. 26, Spring, pp. 122–140.
55. Ibid.
56. Ahel, O. and Lingenau, K. (2020), "Opportunities and Challenges of Digitalization to Improve Access to Education for Sustainable Development in Higher Education", Leal Filho, W., Salvia, A.L., Pretorius, RW, Brandli, LL, Manolas, E ., Alves, F., Azeiteiro, U., Rogers, J., Shiel, C. and Do Paco, A. (Eds), *Universities as Living Labs for Sustainable Development: Supporting the Implementation of the Sustainable Development Goals*, Springer International Publishing, Cham, pp. 341–56.
57. Bayne, S, Evans, P, Ewins, R, Knox, J, Lamb, J, McLeod, O'Shea, C, Ross, J, Sheail, P, Sinclair, C (2020) *The Manifesto for Teaching Online*, Cambridge, MIT Press.
58. Ibid.
59. Leire, C., McCormick, K., Richter, J. L., Arnfalk, P. and Rodhe, H. (2016), "Online teaching going massive: input and outcomes", *Journal of Cleaner Production*, Vol. 123, pp. 230–3.

60. McLean, J. (2019). *Changing digital geographies: Technologies, environments and people*, Springer Nature.
61. Taylor, C (2019) *Posthumanism and Higher Education: Reimagining Pedagogy, Practice and Research*, London, Palgrave.
62. Ibid.
63. Ross, J., Bayne, S., Lamb, J. 2019, 'Critical approaches to valuing digital education: learning with and from the Manifesto for Teaching Online', *Digital Culture & Education*, 11(1), pp. 22–35.
64. Law, J (2004) *After Method: Mess in Social Science Research*, London, Routledge.
65. Deleuze, G and Guattari, F. 1987. *A Thousand Plateaus: Capitalism and Schizophrenia*. London: Athlone Press.
66. See Hillier, J. 2007. *Stretching Beyond the Horizon: a Multiplanar Theory of Spatial Planning and Governance*. Ashgate: Aldershot; Hillier, J. 2011 Strategic Navigation across Multiple Planes: towards a Deleuzean-inspired Methodology for Strategic Spatial Planning *Town Planning Review*, vol. 82: p. 503–527. Hillier, J (2015) Strategic Navigation., *EspacesTemps.net*, Travaux, accessed at https://www.espacestemps.net/articles/strategic-navigation/ on 12/10/2017.
67. See *Sustainable Development Goals* accessed on https://sdgs.un.org/goals

7

What Does Success Look Like?

The Black Box of Transformative Change

Central to the SDGs as a critical framework and agenda for universities is the concept of transformation—deep change that can be forced or chosen, or somewhere in between.[1] To be 'for' transformational change is not to be undiscerning or naïve. Like the concept of resilience discussed in Chap. 3, transformation is far from 'apolitical, inevitable, or universally beneficial'.[2] By virtue of the fact that it can involve painful transitions and has 'the potential to produce significant material and discursive consequences',[3] including involuntary ones, transformation can have a dark side. Yet, when the *status quo* is untenable or under threat, transformation can be the best option. And when the *status quo* is tolerable but far from as good as it could be, transformation can also be the best option.

Both of these situations describe the transformational change called for by the SDGs. The SDG agenda makes clear that transformation is needed to not only redress intolerable inequities and reduce catastrophic risks (risks that threaten to impose swift and deadly transformations if unheeded) but also lift ambitions and enhance the presence of positives such as quality education. The centrality of education to the SDG agenda, as outlined in Chap. 6, is one of the many reasons why the SDG agenda holds such far-reaching implications for higher education. Others include the crucial role of research in SDG problems and action; the ability of universities to facilitate connections between sectors, to work internationally and to influence public discourse; the impact of universities as large investors and real estate managers; and the diversity and size of the groups and perspectives they encompass. All aspects and parts of higher education are implicated in the SDG agenda as both enabler and target of change.

The SDGs have swiftly amplified the importance of universities—and many other institutions and organisations—by underlining the need for the sort of positive social and environmental outcomes they are often uncritically presumed to produce, and by shifting the narrative around sustainable futures from being 'over there' to something everybody needs to create *here/now*. At the same time, the SDGs call universities and its individuals to account. More than just continuing our existing good work, those of us in universities are called upon to acknowledge our role in perpetuating problems and slowing progress, whether by sustaining barriers to integration and inclusion or remaining invested in aggressively unsustainable development. To contribute to transformational change in society we need to provoke the same kinds of changes in higher education.[4]

The case of universities makes clear that it is not only which organisations or sectors are involved in implementing the SDG agenda, it is how—in what ways and to what effect—are they are involved. Universities are already active on the SDGs but in many cases not to the extent or in the ways that are most needed. In a sense, it has been too easy for them. Universities are a natural fit with the SDG agenda thanks to their established expert role, contemporary interest in research impact, and long-standing commitment to the public nature, role and contribution of higher education. While this deeply reciprocal relationship between universities and the SDGs can lead to complacency, it also radically heightens the implications of the SDGs for universities. It opens up numerous 'lines of flight' for linking academic service/scholarship/advocacy/activism to broader societal and political futures and imaginaries to help shape here-to-fore unknown possibilities.

Fournier describes these lines of flight as the elusive moments when change happens when cracks in the often tightly controlled and circumscribed status quo open new spaces of critique and opportunity.[5] The future outcomes of such change are often unclear because 'lines of flight are not headed on any particular trajectory'.[6] They are instead beginnings and possibilities, future-oriented but not reductively so, constantly circling back to reappraise the past and ever-changing present.

One of the cracks that are now opening up as a result of the SDG agenda are hard questions about the role of universities in sustainable development—past, present and future. As Boaventura de Sousa Santos argues, universities are 'undergoing—as much as the rest of contemporary societies—a period of paradigmatic transition ... and it's as important to look back as to look forward'.[7] This refers to the need to look: (1) backwards to understand how growth and development have brought largely unrecognised costs as well as opportunities; (2) at the present to see how collective action is beginning to, or could navigate and mitigate the effects of unsustainable growth; and (3) envisage a future in which sustainable development permeates all elements of the university including its leadership, daily practices, culture and overall impact (see Fig. 7.1).[8] Attention to these three interrelated time frames enables universities—and by extension, wider society—to address the crucial challenge of how to acknowledge, repair and avoid repeating the mistakes of the past. That is—it allows them to learn and transform.

In this chapter we explore how universities can face this challenge by revisiting the matrix framework for transformative change in higher education we introduced in Chap. 1. In particular, we juxtapose the dominant model of university SDG success based on metrics and indicators

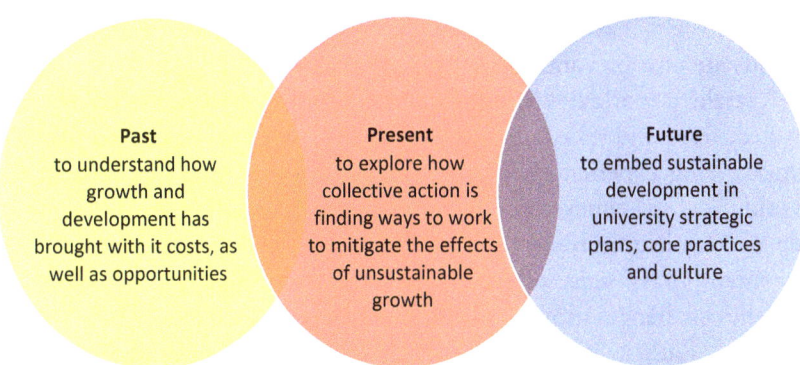

Fig. 7.1 Universities and the SDGs—past, present and future

with growing calls within and beyond universities to re-imagine what success looks like, including success linked to SDG action. We argue that beyond adding a list of new metrics, the SDG agenda in higher education needs to be framed as part of a broader quest for 'the good university' that seeks to build sustainable and just societies that are able to co-exist within a flourishing, healthy planet. We conclude by highlighting the need for intentionality in relation to transformative change as critical to what notions of success might look like—not as a blueprint plan, but as a deep-learning process for universities around 'becoming sustainable'.

Shifting from Disengagement

All institutions and groups are inevitably helping shape the implementation of the global SDG agenda, even if they choose to do nothing (i.e. remain disengaged). The question is not so much whether or not they actively engage, but *how* they do so and whether they can do more to help cultivate the SDGs transformative potential. Empty, self-serving engagement with the SDG agenda is one reason the agenda is dismissed by some groups as merely rhetorical, or what Ruth Levitas would call a 'compensatory utopia'—an image of the future designed to 'educate desire' and guide critiques of the present, but not actually generate change. In contrast, we argue, the SDG agenda offers a practical utopia, one 'intended as goals, as real projects'.[9] Such utopias in action rely on three prerequisites: awareness that society *can* be changed by human agency (a transformative insight itself, as discussed above); a belief that progress or better worlds are possible; and an absence of fatalism.[10]

A fatalistic worldview is one in which people feel society is rigidly controlled, fixed and selfish. It can be a natural outcome of profoundly disempowering experiences, of the sort common to many groups in the world. Yet, as Levitas argues, the widespread, arguably growing, prevalence of fatalism in society is a barrier to practical utopian action. While a contemporary sense of the world being in decline (environmentally, socially) challenges utopian projects, it 'is not in itself an obstacle to utopianism'. Rather, 'it is fatalism that is the key issue' because in denying that society can be profoundly altered, 'much of the motive for the construction of utopias as goals is lost' and so too is their 'transformative element'.[11] The key, Levitas concludes, is cultivating a hope and belief in our own individual and collective agency so that the transition appears practically possible. This is not the job of the utopian vision itself but of the fields of action where change is needed. Hope needs to be 'invested in an agency capable of transformation'.[12]

The uneven distribution of agency across society is not merely a matter of mental outlook or 'attitude', it is itself a reflection of the deeply inequitable structures and systems that the SDGs are trying to improve. While individuals' sense of agency and actual power is far from determined by their social position, some groups are privileged with more influence over our social systems than others. Mediating individuals' agency for better or worse are institutions. While unfashionable, institutions enable as well as constrain our agency, they 'hold us in time and they connect us to each other … they are part of explaining what has gone wrong, and central to working out what we might do to make it right'.[13]

Universities can be agency-generating institutions thanks to the platform, gateway and resources they offer. Moreover, universities believe in and symbolise human agency and social improvement—this is their core developmental nature. Although this aspirational element has been perverted by capitalism, as Tamson Pietsch notes: 'Universities still work with an understanding of time and human capacity that stretches beyond the frames of annual reports, funding cycles, government elections or even of individual careers'.[14] For this reason, universities are a vital field of action for working on utopian projects such as the SDGs.

Although many of us within universities frequently feel despair and perhaps fatalistic about our own agency, as institutions based on a developmental vision of a better future, universities do have a special capacity to demonstrate that desired futures can be generated through practical action. In doing so, they can help cultivate a wider belief that such change is possible, helping enable the SDG project in a profound, indirect way. To play this wider catalytic role, however, they need to ensure that they help expand and diversify the field of action for SDG engagement. They need to work with a wide range of groups within and beyond their walls in an equitable and encouraging way, ensuring that their conservative and elitist tendencies do not perpetuate the sort of heavy bureaucratic feel that threatens to lock the utopian vision of the SDGs into the realm of empty rhetoric and cynicism.

In Chap. 1 we outlined four possible scenarios for how universities might engage with the SDGs and our arguments for why universities should take the option of transformative change (i.e. Deep, Ethical and Bold Engagement) seriously (see Fig. 7.2). The scenarios are structured around the two axes of *commitment* (from shallow to deep) and *ethical innovation* (from conventional to bold). Together they provide a useful heuristic tool for thinking through options for the university and their implications. In particular, they prompt reflection around two key questions: How deeply will the university commit to the SDGs—now and into the future? How bold and ethical will the innovation culture be—in what areas, why, when and by whom? We return to consider these scenarios in the following sections of the chapter.

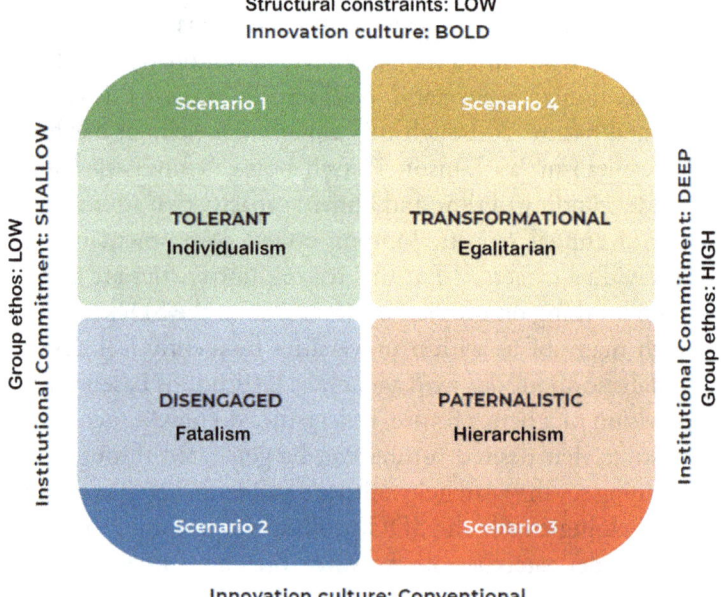

Fig. 7.2 Four possible scenarios for university engagement with the SDGs, overlaid with the four worldviews of Cultural Theory

7 What Does Success Look Like? 211

There is an important alignment here between these scenarios of SDG engagement and the Cultural Theory framework that distinguishes four worldviews on the basis of two similar axes: group ethos and degree of structural order (referred to in the theory as how 'grid-like' society is) as outlined in Fig. 7.2. A strong group ethos and sense of moral commitment to others encourage in institutions a strong commitment to a collective goal, particularly one oriented at assisting others. Conversely, a weak group ethos encourages only shallow institutional commitment. A belief in or sense of strongly structured social order encourages standardisation and compliance and thus a conventional innovation culture, while a sense of low structural constraints encourages a culture of bold innovation. What this means is that which scenario a given university gravitates towards in its engagement with the SDGs likely reflects which underlying worldview is dominant within it. Critically reflecting on such worldviews and becoming conscious and adept at shifting between them is a powerful way of understanding and positively shifting systems.

To recap from Chap. 1 and explain the link to worldviews, the details of these four scenarios are outlined below:

- *Disengaged:* The first scenario represents a shallow institutional commitment and conventional innovation culture. In this situation, a university may initiate work on the SDGs, but it stagnates and fades over time, withering away to become just one of a number of reporting requirements and past enthusiasms. Individuals striving to do things differently are implicitly discouraged and will likely move on to other more open-minded institutions or develop a sense of fatalism about their institution's and the university sector's contribution to the SDGs. If many universities adopt a Disengaged stance, their inherent potential to counter fatalism by building not just human capabilities but faith in them will be quickly eroded, allowing the SDG agenda overall to be more easily dismissed as merely wishful thinking.
- *Tolerant:* The second scenario combines a shallow institutional commitment with a bold innovation culture. The tolerance pathway frames the SDGs as a specialist topic that some staff, students and partners are interested in and may be doing creative and important things with. But those actively working on the SDGs are largely left to their own

devices in keeping with an institutional culture of competitive individualism. To the extent they succeed, it is despite not because of the institution, likely nurtured by niches of radical innovation that at least partially exceed the university (e.g. bold experiments with like-minded partners in government, business and community). The tolerance afforded SDG work stems not from a belief in or even understanding of the SDGs but rather from a blanket belief in any and all innovation, as well as an institutional structure underpinned by an individualistic worldview.

- *Paternalism:* In the third scenario a deeper institutional commitment is combined with a conventional innovation culture. In this scenario the university works to embed the SDG agenda as a strategic priority from the top down across its four core functions of research, education, governance and operations, and external leadership. It takes the SDG agenda seriously as a moral obligation, reflecting an inherent group ethos both in terms of a sense of the institution as a single and special entity that ought to have a coordinated approach, and in terms of performing the university's role as a benevolent, elite institution within society. However, this engagement with the SDG agenda may be driven by a desire to (be seen to) be doing the right thing as a responsible institution more than a deep belief in the need to, or possibility of, revitalising sustainable development as a transformative agenda per se. A paternalistic approach to the SDGs reflects an underlying Hierarchical worldview of the sort both universities and SDGs are renown for in some quarters.
- *Transformational:* The final scenario—and the one we support in this book—combines a deep institutional commitment to SDG action with a bold, ethical innovation culture willing and able to drive transformational change. It aims to rapidly transition the university and wider world onto a more sustainable, socially just pathway. This institutional commitment is deep, bold and pioneering, showcasing and sharing different epistemological understandings and pedagogical practices, underpinned by visionary leadership, resources and support. The Transformational approach reflects and cultivates an underlying Egalitarian worldview that appreciates the multiple, dynamic, far-reaching relationships that make up the world and universities' multi-

dimensional role within it. It is a scenario that exposes and calls out the stifling effects of more fatalistic, individualistic and hierarchical outlooks, underlining the importance of worldviews to the sort of systems change and 'attentive innovation' that the SDGs require.

These scenarios are not prescriptive and within each of the four scenarios there are multiple opportunities for taking pathways towards embedding a transformative SDGs agenda. As an internal agenda, the identification and calling out of *institutional disengagement, tolerance or paternalism*, for example, may be an important impetus for *shifting the status quo* from apathy and stagnation, ad hoc activities or tokenistic SDG-flavoured activities. Given the deadening effect of fatalism on implementation of utopian visions such as the SDGs, we especially want to underline the dangers of a Disengaged stance to the SDGs. For institutions (and individuals) in the Disengaged scenario, evolution towards a Paternalistic or Tolerant approach can be an important step forward. Both of the latter encourage positive actions of the sort we describe and can therefore help transition an institution towards a more Transformational style of engagement—one that engages with the SDGs deeply, ethically and boldly as the institution works with its wider community to tackle the 'big' issues, cultivating internal transformations as a catalyst for greater external impact, engagement and change.

For universities to build a transformative agenda around the SDGs, they need to nurture niche initiatives while fostering wider change, working with key decision makers to embed the SDGs in structures and processes, and create synergies across these. Guiding the sort of reflective approach needed are, we suggest, three broad principles for addressing the SDGs:

- *Reframe the agenda*—critically engaging with the transformative agenda of the SDGs, working towards positive impact and engagement that shifts the status quo;
- *Remake the matrix*—looking past obvious isolated cases of positive pro-SDG initiatives to considering the rest (main body) of what the University does and asking difficult but vital questions about how it is contributing to unsustainable development;

- *Nurture niches*—encouraging, enabling and protecting the incubation of new ideas, approaches and publics.

How successful a university is in these endeavours is, then, a key question. So too is the related question of how a university measures and evaluates its success. There are a number of ways in which universities might assess their success with the SDGs—formally or informally; quantitatively or qualitatively; across the institution or specific to a particular course, project or programme. The list is endless and reflects the way success is defined and negotiated within specific institutional settings and communities. This reflects, in turn, what sort of worldview is dominant. Different scenarios of university SDG engagement (e.g. Paternalism, Tolerance) are likely to define and measure success very differently. In the subsequent sections we illustrate this through cases of existing and possible techniques, such as Voluntary University Reviews (VURs).

Towards Paternalism

To move beyond Disengagement, one of the most common approaches is to begin to engage with the SDGs 'formally'. Besides officially signing on to the SDG agenda via UN processes, this includes engagement with the particular goals, targets and indicators the agenda lays out. In accordance with the good governance approach encouraged by SDG 17 on partnerships and implementation, this involves monitoring and evaluating progress. A wide range of data has become a proxy for success in progressing towards the goals and a consequent guide for decision-making about the ongoing allocation of resources and investments.

In situations such as the SDGs where there is a plethora of data that could be collected, it can be useful to use an indicator framework—a 'set of rules for gathering and organising data so they can be assigned meaning'.[15] Unsurprisingly, the formal SDG agenda proposes a specific set of indicators (231 of them) to guide data collection about progress towards its goals and targets. It is important to note two things about this turn to data and indicators. First of all, it predates the SDGs by a long way. Notably, modern universities are *already* strongly governed by metrics and saturated with indicators, reflecting the roll out of a neoliberal New Public Management governance style since the 1970s to make institutions of all sorts more business-like. Combined with the plethora of metrics associated with the research excellence agenda, this means that universities now use an array of indicator frameworks to establish empirically based assessments and guide decision-making, notably investment decisions.

Metric-based approaches to governance can be valuable. For instance, advantages of using indicators include compiling baseline data around a particular topic; improving decision-making processes and current practices; and enabling changes within communities to be tracked over time.[16] However, technical or administrative methods such as indicators are never 'innocent or purely technical'.[17] They are infused with values, assumptions and biases and can create powerful unintended and/or unjust outcomes. In universities, the rise of an 'audit culture' has been met with fierce criticism from many academics who have pointed out that the 'mania for constant assessment'[18] has created perverse effects and the entrenchment of managerial power at the expense of academic freedom, trust and collegiality. Russell Craig and colleagues assert that 'audit-based university performance management systems' have a 'psychotic potential', perversely rendering 'much academic effort less effective'.[19] Critical reflection and care is thus needed in using indicators, even or especially when they seem highly mundane, standard and commonsensical.

Indicators vary in nature and type and there is no universal model of what constitutes a 'good' indicator. They evolve from different disciplines that tend to 'approach the problems of measurement and tracking from different perspectives'.[20] For example, indicators could be strictly quantitative and based on measurable data sources. In the case of learning and teaching, indicators could be developed by studying the number of programmes and courses that specifically mention or address a topic through course curriculum records. They can also be qualitative and based on student and staff perceptions of an issue. For example, a curriculum mapping exercise could be complemented with insights about people's perceptions of content and its applicability to real-world contexts.

Developing indicators that are meaningful and useful is not easy. Research and guidance on the rigorous development, practical application and monitoring of indicator frameworks is still evolving.[21] A mixed methods approach to indicators that combines social observation and multiple sources of secondary data is often encouraged. Coulton and Korbin argue that irrespective of the type of indicator used, they must be able to be calculated or assessed with reasonable accuracy, and the data must be easily available and cost-effective. Importantly, they suggest that indicators 'have to be practical and should have implications for action— whether it is to drive change or preserve the status quo'.[22]

Traditionally indicators have been divided into three quite different types: economic, environmental and social. Economic indicators have been the most dominant and have typically addressed national elements such as employment, production, growth and inflation.[23] Environmental indicators refer predominantly to elements that relate to ecosystem processes and functions such as water, energy and the assessment of environmental impacts.[24] Social indicators have emerged more recently to assess social conditions and changes as well as shifts in urban conditions. Social indicators are often tied to notions of wellbeing for both individuals and society and these indicators have proven to be more difficult to develop and measure wellbeing directly, given how tricky it is to 'translate or operationalise abstract concepts (e.g. health, safety) into measurable terms'.[25]

Integrated indicators are those that do not fall neatly into the conventional economic, social or environmental categories. 'Sustainability', 'healthy cities' and 'quality of life' have evolved as integrated indicators. These indicators attempt to address the complex nature of their subject matter. Due to their very nature, these socially orientated indicators raise ideological and ethical issues around their role and usage, as well as their relationship to the real world. Their development thus requires a transparent understanding of the conceptual models and underlying theories that have guided the translation of the abstract into something more concrete.

International development has been far from immune to the 'new world order' of audit culture.[26] Monitoring and evaluation of development projects is now a professional field in its own right. Ironically, however, it is in this world of randomised controlled trials and globalised indicators that the politics and partiality of indicators have become especially apparent, reflecting and intensifying the wider politicisation of development.

A critical review of targets for the Millennium Development Goals (MDGs) outlined how indicators act 'as a technology of governance' and able to exert powerful influence by: (1) setting performance standards against which progress can be monitored, rewarded or penalised, and (2) creating a 'knowledge effect' where the indicators intended to reflect a concept effectively act to redefine it.[27] The review concluded that there have been 'many unfortunate, largely unintended, consequences of *simplification* which framed development as a process of delivering concrete and measurable outcomes'.[28] These included:

- Diverting attention from important objectives and challenges
- Creating a silo effect
- Providing unintended incentives by setting the bar too low
- Designing indicators that were conceptually narrow, vertically structured and heavily reliant on technological solutions, neglecting the need for social change
- Framing the concept of development as a set of basic need outcomes, rather than as a process of transformative change in economic, social and political structures.[29]

More broadly, as Diana Liverman points out, the normalisation of indicators in international development work is argued to legitimate neoliberal processes and calculative practices, and bias development investments towards those targets thought to be the most amenable to measurement.[30]

The strong emphasis on indicators in the SDGs framework supports a culture of 'governance by indicators', accelerating trends in quantification and the use of 'governing by goals' to steer the production of evidence and knowledge for policy. This includes the use of SDG-based indicators among universities that are arguably predisposed towards governance by data. The related indicator-based approach to the SDGs resonates with the Paternalistic engagement scenario we outlined above. While it does offer subversive potential (as we discuss below), it is also one that demands ongoing care and vigilance within the context of higher education.

Reorienting Indicators of University Performance Towards the SDGs

The use of indicators to embed the SDGs into higher education is a newly emerging agenda that has been gaining momentum as universities around the world formally commit to advancing the SDGs (see SDSN Australia/Pacific 2017; GUNI 2018; HESI 2019). An example is the *Proposal of indicators to embed the SDGs into Institutional Quality Assessment* by the Quality Assurance Agency for Higher Education of Andorra (AQUA) in collaboration with the Aragon Agency for Quality Assurance and Strategic Foresight in Higher Education (ACPUA), undertaken as part of the project 'Making connections between the institutional evaluation and the sustainable development goals'.[31] The aim of the proposal is to develop a whole-of-institution dialogue and strategic approach to connect an institution's quality assurance framework with the SDGs across all aspects of higher education. The framework builds on lessons learnt from quality assessment around Education for Sustainable Development (ESD) initiatives and is designed to assist universities with:

- *Interpreting* the SDGs in their higher education context;
- *Identifying* quality concerns that are relational to embedding the SDGs in higher education; and
- *Developing* indicators that could be used to improve, as well as assess, an institution's quality performance.

The framework is not confined specifically to the SDGs but focuses more broadly on sustainability and sustainable development and how this relates to the SDGs transformative agenda. The process is intended to be *collaborative,* focused on leadership and management, learning and teaching, research and knowledge exchange, staff and student experience, campus management, partnerships and outreach. A selected summary of one component of the indicator framework—Governance and Strategy—provides a guide for the focus and is outlined in Table 7.1.

In seeking to embed the SDGs into the institutional quality assessment of universities, the AQUA_ACPUA proposal seeks to move beyond

Table 7.1 Embedding the SDGs in higher education through an indicator framework—example of AQUA and ACPUA's governance and strategy framework

Components	Indicators	Assessment criteria
Governance and strategy	1.1 The SDGs form part of the institution's governance framework and implementation is reported in a transparent manner.	Evidence is submitted to confirm that: (a) The University Council or Senate has explicitly committed to sustainability and the SDGs (b) The Executive has explicitly committed to Sustainability and the SDGs
	1.2 The SDGs are included in university strategic documents as well as the University's planning cycle.	Evidence is submitted to confirm that: (a) The strategic framework or plan of the university recognises the SDGs (b) SDGs are embedded in the planning cycle (c) SDGs are embedded in the targets of the strategic framework or plan
	1.3 The implementation of SDGs is monitored and evaluated in line with targets and outcomes identified in the strategic documents.	Evidence is submitted to confirm that: (a) There is monitoring and evaluation in place (b) The outcomes of the evaluation inform the strategic work of the University
	1.4 Leading practice in implementing SDGs is recognised through internal and external awards.	Evidence is submitted to confirm that: (a) Staff have been recognised internally with a certificate/prize/seed funding, promotion (b) Leading practice examples have been recognised by an external award scheme or similar

compliance to include stakeholder participation in the development and implementation of initiatives. The intention is to recognise that an SDGs indicator framework is not a linear but a reflexive and circular process: one that fosters learning and innovation rather than compliance. A review of the initiative concludes:

> Those leading the project were concerned that the SDGs could result in compartmentalization of sustainability and superficial exploration, as many would be tempted to limit their engagement to an audit or tick-box exercise. However, the project experience has shown how the SDGs have acted as doorways eliciting interest in sustainability, originally via thematic pathways that look familiar and interesting to participants, giving value and recognition to existing efforts, but which then join up with other thematic concerns (or objectives) to construct an integrated or holistic framework for sustainability. In the stakeholders' own words, the project had 'shone a light on new pathways' and ignited 'a flame of interest' amongst stakeholders.[32]

Despite the increasing uptake of indicators and goals as success measures for the SDGs, their definition, measurability and outcomes remain highly contentious and tricky. Maria Kaika powerfully argues that, if approached as techno-managerialism, indicators can act as a form of societal immunology, neutralising the potential for real change and discouraging the sort of 'dissensus practices that act as living indicators of what urgently needs to be addressed'.[33]

We agree that indicators can be dangerous and vastly insufficient on their own. However, it is important not to mistake the question of whether to govern university action on the SDGs using indicators with the question of whether universities should be governed by indicators. The latter is already the case and ultimately we may wish to overcome this completely. In the meantime, SDG-based indicators offer a way of decentring *existing* indicators—such as those that make visible only research publications, journal rankings, research income and the number of students. SDG-based indicators can push universities to look past such artefacts to the better world envisaged by the SDG agenda.

In other words, while the question of *whether* universities should be so heavily governed by data remains an important one—given that they are, then using new more progressive indicators can in certain circumstances be a step in the right direction, especially if it helps shift an institution from the Disengaged space or helps generate support for the otherwise neglected hard work on SDGs by some university members. AQUA and ACPUA's approach demonstrates that an indicator framework does not have to be an end in itself but can be a way of flagging issues of concern, in this case, progress around the SDGs.

The highly mainstream new *Times Higher Education Impact Rankings* also illustrates some of the subversive potential of SDG-based indicators. University ranking processes exemplify and drive the hierarchical, competitive developmentality that now characterises higher education. The Impact Rankings are similarly a hierarchical global performance table. In contrast to other rankings, though, this one assesses universities against the SDGs using indicators across three broad areas: research, outreach and stewardship.

- **Research**: to what extent is the university creating knowledge to address the world's problems?
- **Stewardship**: to what extent is the university managing resources and teaching well, and enacting the 'good' university?
- **Outreach**: to what extent is the university directly acting in society to help meet the SDGs?

By introducing a qualitative distinction between research that is, and is not, 'creating knowledge to address the world's problems', the Index exposes the normal agnosticism of the research productivity and research excellence agendas on the core questions of what research is on and for. It disrupts the normal, purely quantitative assessment of research value by introducing the SDGs as an evaluative lens. Although the actual way in which the research is determined to be 'addressing the world's problems' remains limited and highly reductionist (based on key word searches), the Index introduces a range of new considerations. By emphasising the use of open questions that prompt critical reflection around the role of the 'good university', and the extent to which the university has the capacity and intention to meet societal needs, addressing the SDGs are framed and positioned as far more than an area of expertise—but also internal processes and ethos. Of particular note is the emphasis on universities' management and sharing of resources, reflecting wider moves to make organisations accountable for 'externalities' such as greenhouse gas emissions and water use, and more generous towards their local communities and other constituents.

Again, there is significant room for improvement, such as the need to attend to how universities influence the world as large financial investors who may (or may *not*) choose to invest ethically. The point is that the new Impact Index opens the way to include the 'inner workings' of universities as physical entities in and of the world. This is a radical break from the conventional image of universities as largely disembodied nodes of knowledge production and financial flows, which underpins the standard university rankings cultivate. The Index's emphasis on different forms of outreach—not limited to typical knowledge dissemination and public education but including sharing campus facilities and services with community, for example—further supports the necessary new framing of universities as *in* and ideally *of* the world. Thus while governance by data remains problematic, they do offer a window of opportunity to ask critical questions and make visible previously neglected absences, as part of the broader SDG agenda.

Towards and Beyond Tolerance

In addition to, or instead of engaging with the SDG agenda formally, through an expansion and adaptation of their existing audit cultures, some universities are engaging with the SDGs in a more bottom-up manner, demonstrating the potential for more holistic and collaborative approaches to engaging with the SDGs. From an institutional perspective, these are generally in keeping with what we call a Tolerance scenario. However, given that our framework is a heuristic tool only, most examples of existing university approaches are hybrids of the ideal types we outline in our scenario framework.

One example is the experimental indicators that have emerged through the UN's voluntary review process at both the National (VNR) and Local scale (VLR). VLRs monitor the progress of *local* actors towards the achievement of the SDGs across each of the 17 Goals and the associated targets. While this can seem top down, the crucial point is that they use *locally developed* indicators and often *locally collected* data for benchmarking themselves and monitoring specific needs and challenges. According to the United Nations Development Group, localisation refers to:

> the process of defining, implementing and monitoring strategies at the local level for achieving global, national and subnational sustainable development goals and targets This involves concrete mechanisms, tools, innovations, platforms and processes to effectively translate the development agenda into results at the local level. The concept should therefore be understood holistically, beyond the institutions of local governments, to include all local actors through a territorial approach that includes civil society, traditional leaders, religious organizations, academia, the private sector and others.[34]

A VLR is a tool that was originally designed to allow cities and local councils to assess their achievement of the SDGs and their contribution to the 2030 Agenda. It enables cities to prioritise actions and raise awareness about sustainability both within the administration and within the local community. New York City was one of the early adopters of an SDGs VLR, which it publicly presented in 2018. Many other municipal governments have followed suit. Across those involved, reported benefits of the VLR process for cities include:

- *Internal benefits for the city*—cultivating hidden connections, common framework, links between priority and data, sustainable networks;
- *External benefits at local scale*—encouraging transparent accountability, new cross-sectoral partnerships, building leadership;
- *External benefits at global scale*—engaging with the global community and elevating city leaders and priorities within the global conversation.[35]

As a holistic process, the VLR is as much about the journey as the destination. Local partnerships and networks are a central feature of the SDGs framework and the VLR process offers opportunities to strengthen links and foster collaboration both internally with students and staff, and externally with the community and other stakeholders. A number of steps have been identified to assist with the preparation of VLRs and involve a wide range of actors and the collection of different types of data (both qualitative and quantitative) to form an integrated review profile and database. This may involve the identification and mapping of strategic goals against the SDGs at scale.

The VLR process aims to create a pathway for transformative action by identifying not only strengths but the silences that require urgent attention. By identifying priorities and ways to better address the SDGs, the VLRs can raise awareness, map activities across diverse areas and engage diverse stakeholders. Within the city context, the VLR seeks to be accountable, replicable and affordable, trackable over time (at least every three years), rooted in verifiable data analysis and comparable with other cities. Lessons learnt by cities who have undertaken the VLR highlight the opportunities and challenges of such an approach, including the fact that they can accommodate different styles and *vary in scope* to include a review of all SDGs or a selection of SDGs.[36]

Given its success with cities, the VLR has been adapted for Universities as a *Voluntary University Review* (VUR), generating interest from universities in Australia, the EU and South Africa. Carnegie Mellon University (CMU) in the US is the first University globally to publicly commit to—and report on—SDGs VUR. CMU is a small privately funded, research university with programmes in science, technology and business, to public policy, the humanities and the arts. The SDGs align with the University mission and the motivation for the VUR is to 'create a transformative educational experience for students focused on deep disciplinary knowledge; problem-solving; leadership, communication, and interpersonal skills; and personal health and well-being' by 'creating and implementing solutions for real problems, interdisciplinary collaboration, and innovation'.[37]

To initiate the SDGs VUR, the CMU undertook a formal commitment through which they: (1) hired an executive fellow; (2) established a web-page and email address for all SDGs-related queries; and (3) initiated a range of activities to engage the CMU community in discussions about the SDGs (e.g. a podcast, articles in the CMU community publication and dissemination of information about SDG-related activities occurring on campus, an interactive SDGs exhibit). This was followed by a *Knowledge, Attitude and Practice survey* to understand the CMU community's existing activities and level of interest in the SDGs. Additional activities included collaborations with key partners such as The Brookings Institution and The Rockefeller Foundation to gather further strategic insight and information on the SDGs at CMU. The VUR was conducted by a Steering Committee, an executive fellow, a project administrator, a research associate, a sustainability intern and students enrolled in a special summer project course. The Advisory Council and wider members of the CMU community were consulted and provided important input.[38]

The VUR adopted by the CMU did not seek to adopt the metrics proposed by the UN to support the Voluntary *National* Reviews (VNRs) on progress towards SDG targets. Instead, its focus was on the SDGs framework as a cross-cutting sustainability agenda and thematic issues identified for each SDG. Rather than using an existing indicator framework such as the Times HE Impact Index to measure progress and success, the VUR process tracked activities based on desk-top mapping, in-person consultations and a review of CMU information submitted to the Association for Advancement of Sustainability in Higher Education (AASHE). In this sense, the VUR arguably illustrates a bottom-up approach, one that demonstrates the way the SDG agenda can be engaged by allowing initiatives to blossom 'naturally'.

It also serves as a stepping-stone to further action. CMU's VUR pointed to various areas that need attention if SDG work at the university is to flourish. These include:

- Putting in place a more systematic or comprehensive process to collect information on CMU's education, research and practice as it relates to the SDGs.
- Increasing awareness across the wider CMU community of how interconnected the SDGs are (e.g. that they address topics such as racial inequality, gender empowerment, safe migration, police violence and many other pressing societal issues—not just the environment and climate change).
- Connecting groups working on specific SDGs in disciplinary silos across the university who were unaware of each other's work.

- Generating new SDGs-related initiatives to respond to COVID-19 and confront racism.
- Strengthening coordination and engagement on the SDGs within the CMU community, within the localities where CMU operates, and with other entities committed to achieving the SDGs.
- Recognising and rewarding SDG work that falls outside conventional coursework or research categories.
- Enabling more student-centred initiatives such as the 'seven summer project' course where students conducted outreach to student organisations to investigate how their activities relate to the SDGs.
- Increasing recognition that part of the value of the VUR is to encourage reflection and increase intentionality in engagement of the SDGs as an organising and inspirational framework.
- Continuing the VUR as an ongoing, iterative, reflexive and flexible process.[39]

In terms of our framework, the VUR has arguably helped move the university from Tolerance towards Transformation. The CMU provost James Garrett has promised to build the SDGs into the CMU's goals, making six public commitments:

1. We commit to educate CMU students around the world about the SDGs, recognising that this framework applies to all of us and represents a special opportunity to create a more peaceful, prosperous planet with just and inclusive societies.
2. We commit to help solve pressing problems brought to light by the SDG framework by acting boldly, taking risks and applying creativity.
3. We commit to do this work collaboratively, an approach deeply embedded in our university culture.
4. We commit that through education, research, partnerships and operational activities, we will demonstrate advancement of the SDGs at CMU.
5. We commit to create a Voluntary University Review of work being done at CMU and will report these findings in New York City as the UN General Assembly meets next year.

6. We therefore commit to do more to align our work with the SDGs and build on the good work already done by CMU faculty, students, staff and alumni—whether focused on mitigating climate change, eliminating food waste, reducing violence or ending human trafficking.[40]

The CMU experience illustrates that a VUR provides a possible pathway for embedding the SDGs within a university. It underscores that the SDGs are a reciprocal agenda for universities, and that they have key roles to play: (1) in working with their communities to lead efforts to achieve the SDGs; and (2) reshaping their own policies and practices in line with the SDGs framework and agenda. Furthermore, the CMU experience sends an important signal that individual and institutional action *can* build on the SDGs framework to create transformative change at scale, demonstrating and helping cultivate a vital sense of agency and thereby helping rescue the SDG agenda from empty rhetoric.

It is important to acknowledge that while the VUR is bottom up in some respects, the CMU VUR example also demonstrates the need for leadership, commitment and resources to turn the SDG agenda into meaningful action. CMU is a small, privately funded university. The capacity for large publicly funded universities to undertake a formal VUR is as yet unknown. Nor is there certainty about the capacity of early adopters such as CMU to continue to advance the SDGs in the contemporary political and institutional environment given that higher education remains predominantly and firmly framed by discourses of economic growth and development. To help positive niches such as CMU's work on the SDGs to thrive, other efforts are needed to help change the higher education landscape.

Towards Transformation

The SDGs agenda with its ethical focus on re-thinking the pathways and goals of sustainable development is, at heart, about 'people in place' and their inseparability from the health and sustainability of the planet. The specific emphasis in the SDGs on the need for 'transformation' raises the stakes by bringing ethical and procedural focus to the centre of sustainable development. This requires attending to two key critical questions broadly outlined by Mark Pelling, Karen O'Brien and David Matyas:

- Who or what processes determine the mode and objects for change in higher education?
- What are the transformative pathways that will allow action on the SDGs to flourish within the university context?[41]

In her critical assessment of 'The Good University' and the need for radical change, Raewyn Connell argues that fragments are already existing, but need to be brought together in the service of society and planet. This is not universities as resilience machines, but as 'the weave of collective responsibility, labour, activity and possible futures'.[42] Alongside the need for universities themselves to be sustainable, Connell argues this brings to forefront the principles of democracy, truthfulness, creativity and engagement in order to serve society (but not always to agree with it). Building the lines of flight within higher education to sustain both humans and non-humans requires taking action to redress injustice and build community by finding different ways of approaching impact or success.

> The post 2015 development agenda and the SDGs need not only to go beyond "finishing the agenda of the MDGs" but also beyond setting goals and targets. Quantitative targets are powerful as a communications tool and can provide benchmarks for monitoring progress. But a transformative future development agenda requires a qualitative statement of objectives, visionary norms and priority action needed to achieve the objectives including legal, policy and global institutional considerations.[43]

Existing approaches within higher education to generate impact are not adequately meeting societal and planetary needs, nor are they meeting societal expectations or building public trust. If academic institutions are to secure their future, they need to demonstrate a genuine commitment and capacity to work with others to achieve the transformational changes needed. Part of this challenge—and opportunity—is to re-imagine the role and nature of what constitutes success. All universities have an impact culture of some kind, even if that culture is to devalue broader societal impact relative to other agendas as a proxy for return on financial investment. Critical understandings and practices around success and impact can and must evolve in ways that better address meaningful real-world change.

The idea of impact is progressing from being just a required compliance add-on, market-based indicator or instrument for academic promotion, towards one critically focused on the values, purpose and 'spirit' of research that seeks to enable progressive (i.e. political) change towards the type of world we need to co-produce and create. To this end we outline what we describe as *3rd Generation* impact which uses the power of critical praxis to call out misrepresentations and abuses of impact and fight for a better future for universities and those they are genuinely meant to serve. A *3rd Generation* impact culture is prefigurative—it is needed *now* to generate its own conditions for flourishing. It is also regenerative. This is about forging a range of strategic alliances and tactics and working in both overt and covert ways to generate better, fairer and more sustainable futures, from the inside out. We refer to this as a transformative ethos.

There are three different stages of impact that we conceptualise as currently co-existing within higher education. As we describe in more detail below *1st Generation* has a focus on academic relevance and investment reciprocity, *2nd Generation* focuses on the role of research partnerships and value-adding embedded networks, and *3rd Generation* involves universities critically engaging with how, in what ways and to what ends notions of success and impact are being imagined and pursued (see Table 7.2).[44]

Table 7.2 Generating impact in higher education

Impact culture	1st Generation	2nd Generation	3rd Generation
Key foci	Demonstrating academic rigour and relevance to encourage end-user uptake as impact	Working more actively to ensure legitimacy and collaboration within impact culture and literacy	Purposefully fostering the co-production of impact across boundaries

1st Generation Impact Culture

Many researchers and related institutions still think of impact as a matter of defending public and private investment in the university research sector. We describe this as *1st Generation* impact where the dominant approaches include encouraging a given group of intended end-users (e.g. manufacturers, policy-makers) to adopt the research and thereby help transform it into an innovation; and translating and disseminating research in the academic and public domains by making it intellectually and practically accessible, akin to university 'outreach'.

These approaches can be useful and important, yet this *1st Generation* approach to impact risks being compliant, formulaic and superficial within the broader institutional and societal context. Researchers can struggle to see the point, and are generally under-resourced to assist, understand or even hear about any audience engagement with their work. 'End-users' may not exist in reality or not appreciate being told to use something they may not really want, while actual commercial beneficiaries may lock innovations behind closed doors, limiting their value to researchers and the world. In the midst of this, research funders can be underwhelmed by inflated claims of impact and rigidly focused on an unrealistic ideal of initiatives producing demonstrable, quantifiable, attributable outcomes.

More fundamentally, a *1st Generation* impact culture leaves unexamined the deeper questions around: *Who* informs research and pedagogy? *Why* do the research/learning and teaching/engagement? *Who* benefits from higher education?

2nd Generation Impact Culture

In response to growing cynicism around the drivers of *1st Generation* impact, we argue there has been a shift towards *2nd Generation* impact. This latter approach recognises that in order to produce positive impactful changes in the world, universities need to appreciate that problems and solutions are not self-evident or only of their choosing. What counts as a real problem or a satisfactory solution for a given set of stakeholders is always contestable, always a matter of shifting priorities and circumstances. This includes greater recognition that the impacts that universities generate are not always or necessarily positive from the perspective of designated 'end users'—but can be maladaptive in ways unintended.

Rather than just seeking to maximise the impact of a given area of research, the goal extends to working collaboratively to generate outcomes that research partners recognise as valuable. Centring partnership perspectives means that *2nd Generation* research impact aims for research that is both relevant and legitimised.[45] The linear 'push' of *1st Generation*— *from* researchers *to* end-users—is replaced with a process that is more circular and iterative, includes social, cultural and environmental priorities in addition to economic and bibliometric ones, and recognises that impact emerges out of relationships and needs to be supported by impact literacy across the university.[46] Yet even as this *2nd Generation* impact seeks to gain momentum, an even more transformative and ambitious approach is needed to drive change.

3rd Generation Impact Culture

A transformative approach to impact focuses on the need for change not just 'out there' in the wider community but also 'in here' within dominant university policies and practices. It asks not just '*what*' new research technologies or data are needed but also '*how, why, for whom and to what ends*' higher education impact is able to support positive societal change. Impact is understood as potentially vital and even radical. It requires critically engaging with both the means and the ends, including the role of an institution's impact culture and what is possible when a positive impact culture is deliberately cultivated.[47]

A *3rd Generation* impact culture cultivates an ethos of impact that doesn't just scale out and up but also aims to *scale deep* through critical engagement with the systemic and societal nature of the societal challenges being faced.[48] It encourages universities and researchers to take the question of impact as a serious question, learning opportunity and critical change agent in its own right. It recognises that research and L&T

involve value-laden decisions from start to finish, and inquires into the collective impact of these and our other endeavours in universities.

A *3rd Generation* impact culture also actively encourages—even requires—bottom-up, community-grounded approaches to reshaping existing hierarchies that inhibit real change. Taking a critical, interdisciplinary approach to the 'public value' of higher education, it calls for difficult conversations in and about universities and related communities of practice. In particular, it calls into question the extent and ways that university-based activity is actually helping the world address key challenges such as the climate emergency. This is why—although a range of impact cultures and success indicators may currently co-exist within universities—it is *3rd Generation* impact that is urgently needed. At an institutional and sectoral scale, it pushes forward the following questions for evaluation:

- What type of world are we helping generate through our universities, individually and collectively?
- What do we need to do more of, or less of, differently?
- How can we create positive impacts across and between the work we and our institutions do?

Overall, dynamic, complex and urgent situations mean that in addition to a diversity of project-based initiatives on the SDGs, larger, more anticipatory, agile, discerning and wide-ranging approaches to higher education success and impact are needed to generate the critical change needed. Although the reach of universities is increasing, so too are impact needs, with current academic practices failing to arrest profoundly dangerous trajectories such as climate change. Too many activities and initiatives remain focused on narrowly defined impacts, too many groups in society remain left out of research conversations, and too often what is asked of researchers by funders is out of step with future challenges and ignorant of realities. But there are alternatives.

Responsible and Intentional Higher Education

The prospect of *intentional* transformational change prompts critical reflection about what we want to change and what we want to protect or grow. Often it necessitates looking beneath surface appearances and revising initial assumptions about goals and presumed inevitabilities. Common to many deliberate transformation efforts is recognition of the non-inevitability of many existing structural patterns and norms. A desire for transformation often pushes community members and decision makers to tackle root social causes such as power imbalances, layered injustices, paradigms, worldviews and values. Conversely, frustration with and desperation about such entrenched problems is often what pushes people to aim for transformational change.

That deep social injustices can and should be redressed is one of the core messages of the SDG agenda. It is transformational in its own right for the way it implicitly exposes society as more malleable than many people assume. Intentional transformational change in society is revealed as a more serious possibility, and the buried choice between normalised incremental change agendas and more systemic transformative ones is brought to the surface.

Concurrently, the SDG agenda underlines that nature (as in the Earth System that we are a part of) is less malleable than assumed by many—or at least by those with an Individualist worldview (see Chap. 2) which presents nature as tough and mouldable. As indicated in Chap. 3, the SDG agenda is consistent with a broad acceptance that nature—including the Earth System as a whole—has limits to the amount of stress and disturbance it can cope with before flipping (transforming) into another state, one far less habitable for today's living beings. Facing this truth reveals unintentional transformational change across all physical and social domains to be a far more serious possibility than usually acknowledged. The problem of immanent, unwanted transformations in the planet and our living conditions once again brings to the surface the buried choice in society between normalised incremental change agendas and the more systemic transformative ones we need.

The challenge is how to avoid unintentional transformational change by embracing intentional transformational change. Unfortunately, how to generate and guide positive transformational change is poorly understood, given previous neglect of the topic. As climate change adaptation scholars Mark Pelling, Karen O'Brien and David Matyas note, many questions demand attention:

> What is the theoretical relationship between transformation, incremental adaptation, stability and resilience, and how might these processes interact? How and where might transformation emerge and spread? In what ways does transformation provoke changes in the approaches taken by researchers and practitioners?[49]

The intellectual and practical challenge of how to stimulate, coordinate and even research positive transformation is at the heart of the SDGs. With its 17 diverse and ambitious goals (eradicating poverty, tackling climate change, creating safe, resilient and sustainable cities, achieving gender equity, among many others), the SDG agenda demands real systemic change but does not articulate how it is to be achieved. The SDGs are a work in progress, a problem statement more than a solution. Vast knowledge gaps remain internationally around how to plan for and implement them, how to monitor and evaluate progress and how to develop the skills and capabilities needed across governments, business, NGOs/civil society and universities to advance transformative change.

Combined with lack of political will, the result is a growing *implementation* gap as the world keeps charging along in the wrong direction, ignoring the warnings of multiple SDG progress reports that, like a GPS map, are tracing the ongoing, and in some cases growing, distance between where we are and where we should be. For example, society's global material footprint (amount of material resources used) increased 17.4% between 2010 and 2017, rising across all categories of materials (metals, non-metal materials, fossil fuels and biomass) from a total of 73.2 to 85.9 billion metric tonnes per year. Partly as a result, climate change and biodiversity loss are worsening, not improving, at the global level.[50]

Although there is increasing support for the SDG agenda, and much action in terms of planning, indicator frameworks, capabilities mapping and SDG badging, too much is stuck in the promotional and marketing sphere and too little is translating into practical action. This leaves us poorly prepared to cope with new problems and 'external' shocks, such as the COVID-19 pandemic that the *Sustainable Development Goals Report 2020* warns has pushed more than 71 million people back into extreme poverty, reversing gains in reducing poverty since 1998.[51]

As part of re-thinking what success looks like in relation to the SDG agenda we would like to offer three key markers of an emerging transformative agenda for universities. Complements to clear and bold action on specific SDGs such as climate action, these cross-cutting areas highlight the different scales, angles and alliances for action that an SDG commitment can engender in universities.

1. *Explicit recognition of Indigenous sovereignty*

As outlined in Chaps. 1 and 2, engaging with the SDG agenda means engaging with the history of (un)sustainable development and universities' ongoing role within it. Inseparable from this history is colonialism. Part of 'bearing witness' and taking responsibility for the harms of these *developmentalitie*s is redressing silences and inaction on Indigenous truths, rights and sovereignty, both within the formal SDG agenda and universities. Indigenous sovereignty and futures are intimately linked to any meaningful notions of success around the SDGs as a transformative agenda. In universities, recognition of Indigenous sovereignty needs to be embedded into all elements of the institution alongside the SDGs.

Part of the challenge is to re-think forms of knowledge production that privilege predominantly western ways of knowing and being over others: where 'knowledge production and everyday relations are informed by European colonial modalities of power and propped up by imperial geopolitics and economic arrangements'.[52] As articulated by Konai Helu Thaman within the context of decolonising Pacific Studies:

> For me, decolonizing Pacific studies is important because (1) it is about acknowledging and recognizing the dominance of western philosophy,

content, and pedagogy in the lives and the education of Pacific peoples; (2) it is about valuing alternative ways of thinking about our world, particularly those rooted in the indigenous cultures of Oceanic peoples; and (3) it is about developing a new philosophy of education that is culturally inclusive and gender sensitive.[53]

A decolonising approach to the SDGs means querying universal claims to knowledge and interrogating how they marginalise and discount places, people and knowledges across the world[54] and working to build Indigenous sovereignty into SDG responses whether such responses involve research on cities or climate change, teaching on innovation and infrastructure, investment in programmes and partnerships, or campus sustainability and equity initiatives. It means advancing the SDGs by following Indigenous people in asking hard questions around economic ideology, progress and sustainability[55] and pushing for more ambitious change within colonial institutions and people. It means facing ongoing tensions between claims of transformative change and the continuing violence of everyday colonialism.[56]

2. A Strong, Empowered Union and Student Movement

The role of a strong empowered staff and student movement is fundamental to the transformation of the higher education sector towards a more sustainable future. In the face of system-wide inequity and an aggressive economics-first mentality, they are the drivers of action-led change within and through their institutions, particularly when those institutions fail to drive such change themselves.

The importance of university trade unions in advocating for the voices and interests of academic staff, including casualised ones, has been more apparent than ever in recent years as staff have been asked to bear the brunt of myriad financial pressures, worsened but not caused by the COVID-19 pandemic. One of the benefits of these efforts has been to highlight the potential for higher education trade unions to help achieve the transformative potential of the SDGs at the local, regional, national and global scale. Indeed, some trade unions are already very involved in the SDGs, working in solidarity with university staff to 'uphold freedom of association, protect social dialogue and collective bargaining, and promote decent work, social protection and the rights of working people'.[57]

Students are similarly showing solidarity through representation in groups that build capacity to tackle inequality and the root causes of the sustainability crises that the SDGs have emerged in response to. The student-led movement for university fossil fuel divestment, for example, demonstrates an awareness of not only the enormous financial wealth that some universities have but of the negative impact that wealth has in the world if not deliberately and carefully directed to positive ends.[58] In loudly calling out government inaction on climate change, student climate protests also make obvious how silent and complacent most university leaders' are on the issue. Students are also beginning to come together over broader sustainable development issues. In 2018, for example, thousands of African students participated in the Africa Students' and Youth Summit 2018 (ASYS) Kigali, Rwanda to contribute towards the SDGs and African Union Agenda 2063.[59]

Given the intelligence and passion of students and staff, some universities are beginning to involve them in not just one-off events but sustained, transparent and genuine institutional efforts. SDG action by staff and students in higher education cannot be bound by institutions. Networks and associations of all kinds—from discipline-based academic groups to professional associations of research managers, from networks of campus managers to student sports clubs—all need to be enrolled in helping reorient the sector towards more sustainable and just futures. At the same time, all need to be asked to reflect on what they are doing, and what they could be doing, to galvanise positive action on the SDGs. To what extent are they inclusive, equitable, environmentally sustainable and working within their spheres of influence for regenerative futures? Alliances across such groups, further deepen their influence.

3. A Well-resourced and Supported Library

A third key area that we see as a vital sign of higher education engagement with the SDGs is the health of the library. The role of libraries in relation to their contribution to the SDGs specifically has been articulated as the six 'P's of libraries and development which reflect both traditional and emerging roles of libraries.[60] These roles include protecting research heritage and presenting it in a way attuned to its tensions and silences; providing research, and research tools, to support staff and students and enhance the quality of their work; empowering staff and students with the skills and knowledge they need to do critical work, such as how to negotiate academic sources in a discerning and just way; providing portals to other services including those designed to support the wellbeing of staff and students; partnering with those working in other parts of the university to generate positive outcomes such as more equitable access to resources or better research impacts; offering platforms for collaboration between staff, students and other groups, serving as community hubs by hosting courses and seminars for example; and producing events and resources to help increase awareness, engagement and positive impact around the SDGs (see Fig. 7.3).

While they do not attract the same attention as the core areas of universities (research, learning and teaching, leadership and external engagement), libraries are at the heart of universities and can act as critical knowledge brokers and conduits for positive change. For example, some libraries are strong advocates for 'open access' and the sharing of knowledge by making resources such as reference collections available to the community.[61] As both physical and virtual spaces that stretch across and beyond universities, encompassing people from diverse age groups and backgrounds, libraries are also often an essential part of a university's infrastructure of care, providing a sense of wellbeing and belonging. Can you imagine a university without its libraries?

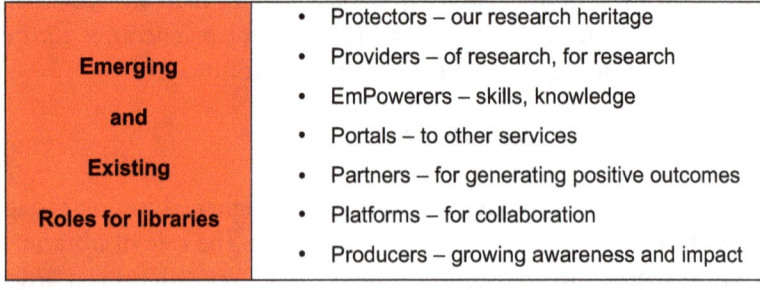

Fig. 7.3 Existing and emerging roles for libraries and the SDGs. (Adapted from IFLA 2020)

Despite their vital role, many university libraries are under budget pressure, especially in the wake of the pandemic,[62] reflecting longer standing struggles to communicate the value of library services.[63] Explicitly positioning libraries at the centre of universities' SDG work could increase the support they receive. To do so, however, requires reconsidering the success of academic libraries in light of SDG outcomes. How do their collections stack up, for instance, from a sustainability, resilience and justice point of view? What sort of world are they implicitly helping create? These are the sorts of questions that all units, areas and services of a university need to ask themselves.

The flourishing (or not) of libraries, staff and student groups, and action on Indigenous sovereignty are all bellwethers of the health and vitality of the university and its capacity, commitment, sincerity and intentionality to really advance the SDGs in a transformative way.

Becoming Sustainable in Higher Education

Success takes many forms and is pursued at different strategic, sectoral, spatial, temporal, virtual and disciplinary scales within contemporary universities. Following Kamola 'It is important to remember that universities are always multiple, with many histories, and many crises'.[64] The unsustainable development trajectory that universities are part of is a critical but neglected element of these crises, one that cannot be separated from or trumped by others. We understand and support critiques of the neoliberalised university model and the crises it has engendered and understand why it has pushed some people to turn away from universities or give up on their transformative potential. However we argue that the importance of the SDG agenda requires that we revive universities, reinstate a more progressive impact imaginary within them and work from the inside out to mobilise transformative change.

It is increasingly apparent that the transformational change demanded by the SDGs requires transformational change in how we work. Three challenges especially need to be tackled. One is the tendency to go for siloed, reductionist solutions. The SDGs are designed (albeit imperfectly) to be an integrative framework, not a menu. Implementing them requires sophisticated, conscious integration, whether by designing activities at the nexus of multiple issues (for food, water and energy), ensuring interventions are implemented in ways that enhance not hinder progress on other goals, forging agreements across different domains about what counts as valid evidence and a feasible proposition, or building innovative, interdisciplinary and transdisciplinary capabilities and practices. All of this requires redoubling efforts to deconstruct and tunnel through the maze of boundaries we have built at multiple scales between different areas of work, including higher education and 'the rest of the world'.

Second, implementation of the SDG agenda needs to be scaled up, out and deep. Many projects identify great potential but are limited in effectiveness because they do not cultivate the enabling conditions needed to ensure that positive initiatives endure and others are more easily instigated. To go beyond a string of isolated and temporary efforts, SDG projects need to scale up into policy and strategy, deep into cultural norms and understanding, and out into new contexts.[65] To achieve this transformation, we need to remedy our over-reliance on short-term, bounded 'projects' and the short-term precarious jobs that go with them—which in itself demands another transformational change.

Third, we need to evolve the collective SDG 'project' from its UN origins and nation-state hierarchy to more *inclusive* decentralised, transnational practices that open the way for community, not-for-profits, business and individuals to contribute more fully through the local and subnational networks, movements and arenas. Unfortunately, many responses to the SDGs agenda to date are fragmented and characterised by indicator-it is, marketing mania and empty elite endorsements (e.g. formal and formulaic statements of support for the SDG agenda by corporations and large government bodies, often completely divorced from staff and everyday activities). Such an approach is not inevitable and better, more inclusive ways of doing things within the university context are possible.

What constitutes success and impact is constantly evolving and will continue to do so as relationships between universities and society shift. 'Transformative' impact will by definition be 'transforming': political and not passive; enabling and not disabling; and with the capacity to disrupt the development *status quo*. An overly rigid or narrow representation of what success around the SDGs looks like and the conflation of impact with measurement of fixed indicators ironically serves to reinforce the infamous moniker—'there is no alternative'. As de Sousa Santos articulates in terms of higher education indicators:

> The weakest of them all are the nonanswers, the silences, and the taken-for-grantedness of the new common sense about the mission of the university.[66]

To generate the *significant impact* needed—no matter the size or scale—requires not only working in inward-facing and outward-facing ways but working across boundaries of all sorts. We need critically engaged cross-disciplinary approaches that link across and weave together impact to challenge, subvert, disrupt, resist, re-imagine, recalibrate society's big challenges and opportunities. Work on the 4th Industrial Revolution, for example, demands the insights of the social sciences and humanities if its impact is to be empowering, rather than divisive and dehumanising. We also need to find the critical synergies and lessons across, within and between projects, programmes, partnerships, networks, associations and institutions to create multiple, co-existing forms of engagement and impact.

Critical understandings of and practices around success need to be part of the reciprocal agenda for universities and the SDGs whose actions and outcomes are mutually shaping and must evolve together in ways that better address meaningful real-world change. 'Becoming sustainable' must be 'worked and reworked as a politics that is already and always in the making'.[67] A vital first step is to help demonstrate that a degree of success is possible, cultivating renewed faith in human agency.

Notes

1. Pelling, M, and O'Brien, K, Matyas, D (2014) Adaptation and Transformation, *Climate Change*, Springer.
2. Blythe, J, Silver, J, Evans, L, Armitage, D, Bennett, N, Moore, M, Morrison, T, Brown, K (2018) The darkside of transformation: Latent risks in contemporary sustainability discourse, *Antipode*, p. 14.

3. Ibid.
4. Pelling, M, and O'Brien, K, Matyas, D (2014).
5. Fournier, M (2014) Lines of Flight, *Transgender Studies Quarterly*, 1(1–2), p. 121–122.
6. Miller, E. (2019). *Reimagining Livelihoods: Life Beyond Economy, Society, and Environment*. Minneapolis: University of Minnesota Press. p. 88.
7. Santos, Boaventura de Sousa (2012) The University at a Crossroads, *Human Architecture: Journal of the Sociology of Self-Knowledge*, 10 (1), Article 3. Available at: http://scholarworks.umb.edu/humanarchitecture/vol10/iss1/3
8. Insights from Troy, P (1999) *Serving the City*, Sydney, Pluto Press.
9. Levitas, R. (2010). *The concept of Utopia*, Peter Lang, London. p. 225.
10. Goodwin, B. and K. Taylor (2009). *The politics of Utopia: A study in theory and practice*, Peter Lang, London.
11. Levitas (2010), p. 226.
12. Ibid., p. 231.
13. Pietsch, T. (2016). On institutions: why we (still) need them. *Griffith Review*, Edition 51: Fixing the System, p. 117.
14. op cit.
15. Innes, J. (1990) *Knowledge and Public Policy: The search for meaningful indicators*, New Jersey: Transaction Publishers, p. 5.
16. Ben-Arieh, A. and Goerge, R. (2006) *Indicators of children's wellbeing: Understanding their role, usage and policy influence*, The Netherlands: Springer.
17. Law, J. (2004). *After Method: Mess in Social Science Research*. New York: Routledge, p. 143.
18. Collini, S. (2010) Browne's gamble. London Review of Books 32, 23–25. p. 24.
19. Craig, R., Amernic, J., Tourish, D. (2014) Perverse Audit Culture and Accountability of the Modern Public University. *Financial Accountability & Management* 30, 1–24.
20. Hoernig, H. and Seasons, M. (2004) Monitoring of Indicators in Local and Regional Planning Practice: Concepts and Issues, *Planning Practice and Research*, 19(1), p. 82.
21. See Alkire S (2007) The missing dimensions of poverty data: introduction to the special issue. *Oxford Development Studies* 35(4): p. 347–359; Barnett C and Parnell S (2016) Ideas, implementation and indicators: epistemologies of the post-2015 urban agenda. *Environment &*

Urbanization 28(1): p. 1–13; Hák T, Janoušková S and Moldan B (2016) Sustainable development goals: a need for relevant indicators. *Ecological Indicators* 60: p. 565–573; Langford M and Winkler I (2014) Muddying the water? Assessing target-based approaches in development cooperation for water and sanitation. *Journal of Human Development and Capabilities* 15(2–3): p. 247–260.
22. Coulton, C. and Korbin, J. (2007) Indicators of child wellbeing through a neighborhood lens, *Social Indicators Research*, 84(3), p. 3.
23. Grant, J. (1999) *A Handbook of Economic Indicators*, University of Toronto Press, Toronto.
24. Muller, F., Hoffman-Kroll, R. and Wiggering, H. (2000) Indicating ecosystem integrity—theoretical concepts and environmental requirements, *Ecological Modelling*, 130, p. 13–23.
25. Hoernig, H. and Seasons, M. (2004) Monitoring of Indicators in Local and Regional Planning Practice: Concepts and Issues, *Planning Practice and Research*, 19(1), p. 84.
26. Shore, C. and S. Wright (2015). Governing by numbers: audit culture, rankings and the new world order. *Social Anthropology* 23(1): 22–28.
27. Fukuda-Parr S, Yamin AE, and Greenstein J (2014) The power of numbers: a critical review of Millennium Development Goal targets for human development and human rights. *Journal of Human Development and Capabilities* 15(2–3), p. 105–117.
28. Ibid.
29. Ibid.
30. Liverman, D (2018) Geographic perspectives on development goals: Constructive engagements and critical perspectives on the MDGs and SDGs, *Dialogues in Human Geography*, 8(2), p. 168–185.
31. la Vella, A (2019) *Proposal of indicators to embed the Sustainable Development Goals into Institutional Quality Assessment,* accessed online at https://www.aqua.ad/system/files/sites/private/files/101_17-045_proposal_of_indicators_to_embed_the_sdg_into_institutional_q_assessment_digital_0.pdf
32. Ibid.
33. Kaika, M (2017) "Don't call me resilient again!" The new urban order as immunology...or what happens when communities refuse to be vaccinated with 'smart cities and indicators, *Environment and Urbanization,* 20 (1), p. 89–102.

34. United Nations Development Group (2016). *Guidelines To Support Country Reporting on the Sustainable Development Goals,* accessed online at https://unsdg.un.org/sites/default/files/Guidelines-to-Support-Country-Reporting-on-SDGs-1.pdf
35. Deininger, Niki, Yasu Lu, Jason Griess, and Robert Santamaria. (2019). *Cities Taking the Lead on the Sustainable Development Goals.* Washington, Brookings Institution.
36. Siragusa A., Vizcaino P., Proietti P., Lavalle C., (202) *European Handbook for SDG Voluntary Local Reviews,* EUR 30067 EN, Publications Office of the European Union, Luxembourg.
37. CMU (2020) *Voluntary University Review of the Sustainable Development Goals,* Carnegie Mellon University Sustainability Initiative, Pittsburgh, accessed online at https://www.cmu.edu/leadership/the-provost/provost-initiatives/cmu-vur-2020
38. Ibid.
39. Ibid.
40. CMU (2019) *Pittsburgh Spotlighted as Leader in Advancing Sustainable Development Goals,* accessed online at https://www.cmu.edu/news/stories/archives/2019/september/sustainable-development-goals.html
41. Pelling, M, and O'Brien, K, Matyas, D (2014) *Adaptation and Transformation,* Climate Change, Springer.
42. Connell, R (2019) *The good university: What universities actually do and why it's time for a radical change,* Melbourne, Monash University Press.
43. Fukuda-Parr S, Yamin AE, and Greenstein J (2014) The power of numbers: a critical review of Millennium Development Goal targets for human development and human rights. *Journal of Human Development and Capabilities* 15(2–3), p. 105–117.
44. Rickards, L, Steele, W, Kokshagina, O and Moraes, O (2020) Research Impact as Ethos, Melbourne, RMIT University, accessed at https://cur.org.au/cms/wp-content/uploads/2020/09/rickards-et-al-2020-research-impact-as-ethos.pdf
45. Cash, D.W., Clark, W.C., Alcock, F., Dickson, N.M., Eckley, N., Guston, D.H., Jäger, J. & Mitchell, R.B. (2003) Knowledge systems for sustainable development, *Proceedings of the National Academy of Sciences of the United States of America,* 100(14), p. 8086–8091.
46. Bayley, J. E. and D. Phipps (2019). "Building the concept of research impact literacy." *Evidence & Policy: A Journal of Research, Debate and Practice* 15 (4), p. 597–606.

47. Reed, M. S. (2016). *The Research Impact Handbook, Fast Track Impact.* https://www.fasttrackimpact.com/books
48. Moore, M.-L., D. Riddell and D. Vocisano (2015). Scaling out, scaling up, scaling deep: strategies of non-profits in advancing systemic social innovation. *Journal of Corporate Citizenship* (58): p. 67–84.
49. Pelling, M, and O'Brien, K, Matyas, D (2014) Adaptation and Transformation, *Climate Change*, Springer.
50. United Nations (2020) *Sustainable Development Goals Report 2020.* United Nations, Geneva.
51. Ibid.
52. Collard R-C, Dempsey J and Sundberg J (2015) A manifesto for abundant futures *Annals of the AAG* 105, p. 323.
53. Thaman, K. H. (2003). Decolonizing Pacific Studies: Indigenous Perspectives, Knowledge, and Wisdom in Higher Education. Special issue, *The Contemporary Pacific* 15 (1): p. 1–17.
54. Radcliffe, S (2017) Decolonizing Geographical Knowledges, *Transactions of the Institute of British Geographers*, 42(3), p. 329–333.
55. Todd, Z.C., (2015). Indigenizing the Anthropocene, in: Davis, H., Turpin, E. (Eds.), *Art in the Anthropocene: Encounters Among Aesthetics, Politics, Environment and Epistemology.* Open Humanities Press, p. 241–254, Whyte, K., 2016. Is it colonial déjà vu? Indigenous peoples and climate injustice; Whyte, K.P., (2018). Indigenous science (fiction) for the Anthropocene: Ancestral dystopias and fantasies of climate change crises. *Environment and Planning E: Nature and Space* 1, 224–242; Williams, S., Doyon, A., (2019). Justice in energy transitions. *Environmental Innovation and Societal Transitions* 31, 144–153.
56. Rose, D.B. (2004) *Reports from a Wild Country: Ethics for decolonisation*, University of New South Wales Press, Sydney.
57. UN (n.d.) Trade unions and the 2030 Agenda, accessed online at https://sustainabledevelopment.un.org/index.php?page=view&nr=4&type=88&menu=156
58. Ibrahim, Z (2020), Universities divesting from fossil fuels have made history but the fight isn't over, *The Guardian*, 13th January, accessed online at https://www.theguardian.com/education/2020/jan/13/universities-divesting-from-fossil-fuels-have-made-history-but-the-fight-isnt-over

59. Ntuli, M (2019) Student activism and its role in achieving the SDGs, *University World News,* 14th December, accessed online at https://www.universityworldnews.com/post.php?story=20191214112317666
60. IFLA (2020) The 6 P's of libraries and the SDGs, accessed on https://blogs.ifla.org/lpa/2020/10/12/the-6-ps-of-libraries-and-the-sdgs/
61. See Hess, C. and E. Ostrom (eds) (2007), *Understanding Knowledge as a Commons*, Cambridge, MA and London: MIT Press; Federici, S. (2009), 'Education and the enclosure of knowledge in the global university', A*CME*, 8 (3), p. 454–61.
62. See, for example, a recent survey of US university libraries. https://sr.ithaka.org/publications/academic-library-strategy-and-budgeting-during-the-covid-19-pandemic/
63. https://www.wiley.com/network/archive/the-top-10-challenges-academic-librarians-face-in-2016
64. Kamola, I (2019) *Making the world global: U.S. Universities and the production of the global imaginary,* London, Duke University Press.
65. Moore, M.-L., D. Riddell and D. Vocisano (2015). Scaling out, scaling up, scaling deep: strategies of non-profits in advancing systemic social innovation. *Journal of Corporate Citizenship* (58): p. 67–84.
66. Santos, Boaventura de Sousa (2012) The University at a Crossroads, *Human Architecture: Journal of the Sociology of Self-Knowledge*, 10 (1), Article 3. Available at: http://scholarworks.umb.edu/humanarchitecture/vol10/iss1/3
67. Shear, Boone W. "Toward an Ontological Politics of Collaborative Entanglement: *Teaching and Learning as Methods Assemblage." Collaborative Anthropologies* 12(1), p. 75.

8

Sustainable Futures

The Urgent Need to Face Injustice and Unsustainability

Is there any other institution (except possibly government) that combines so many social functions? Is ... so diffuse and unreadable in its core objectives? So self-serving and other-serving at the same? So easily annexed to a range of contrary agendas: conservative and radical, capitalist and socialist, elite and democratic, technocratic and organic? ... But the university rarely holds to a single course. It continually disappoints. It always falls short of potential. But we defend it. We sense that if it were lost then something quite fundamental, and probably essential, would be lost.[1]

Sustainable development in the Anthropocene is not about tinkering around the edges. Just as development cannot be genuinely fixed with international development add-ons, sustainability cannot be addressed with green add-ons. Despite all the effort going into devising new 'eco' things—from energy-efficient buildings to electric cars, low carbon clothes to biodiversity-friendly coffee—the gravity of the sustainability crisis demands that we face up to what John Barry calls 'the politics of actually existing unsustainability'. Barry argues that we need to 'identify and reduce existing *unsustainabilities* as a precondition for, and prior to, any aim to articulate and achieve future sustainability or some future sustainable development path'. This means recognising that 'reducing actually existing unsustainability may be as much about "letting go" or reducing existing practices as proposing something new'.[2]

Partly because they have been so environmentally unsustainable, development paths to date have also been profoundly unjust, causing the death, degradation and displacement of people and non-humans around the world. Like unsustainability, addressing this 'actually existing injustice' is also a precondition for future sustainable development and requires far more than add-ons by universities and others. It means acknowledging, arresting and preventing the ongoing social harms perpetuated by dominant systems, and redressing the way in which harm to certain groups is normalised, disregarded and denied by pushing ourselves to design wiser, regenerative approaches that enhance the wellbeing of all.

The SDG agenda aligns strongly with the need to face the politics of actually existing unsustainability and injustice. Encompassing virtually all human activities, many of the goals are expressed as reducing undesirable practices. Some move beyond symptoms to address causes, such as (un)responsible consumption and production, (non)clean energy and (un)sustainable food systems. In this way, the SDG agenda implicitly communicates Aristotle's point that 'What it lies in our power to do, it lies in our power not to do'.[3] That said, the way in which the agenda is being implemented suggests that the politics of actually existing unsustainability and injustice are being side-stepped. Too often it seems that SDGs are being employed only rhetorically or cherry picked and placed alongside business-as-usual activities as a novel side-interest or compensatory marketing-oriented effort.[4]

Our approach in this book has been to position the SDGs as a witness statement to the unsustainable and unjust trajectory of development (including in higher education), and the transformative prospects and pathways for a sustainable future. Combined with the transformational character of the change in the world that the SDG agenda is seeking, this means the adoption of the SDGs in higher education promises to have deep and wide effects for the sector. None of this will be automatic, however. The SDGs require conscious and reciprocal processes of transformative 'change in education' *and* 'education for change'.[5]

Our starting point is the crisis state of the world, and the need to fundamentally reframe the dominant 'developmentalities'. The goal here is to shift attention from a focus on the 'the what' to 'the how' and 'the why' the SDGs are a priority for re-imagining higher education. As we have described, the story of the SDGs agenda is also the story of development. What the agenda does in practice, however, is far from certain. The SDGs represent the goal posts we jointly need to orient towards in the Anthropocene. These goal posts are wide and diverse but represent a significant shift for both universities and society. Moving beyond the nationalistic and individualistic competitor mindset, the SDGs encourage universities to work with others to heed the global call to action.

Like others, we believe that universities are vital to progressing the SDG agenda and have a fundamental role to play across all four of their functions: teaching and learning, research impact, external leadership and internal operations.[6] What we particularly emphasise is that for universities to perform their unique function as enablers of change, they need to simultaneously embrace their role as *targets* for change and ensure they are role modelling the sort of approaches and impacts they want to engender. The urgency and complexity of sustainable development, combined with universities' multidimensional and influential role in creating the present and future, means that they need to be more thoughtful *and* energetic in generating change.

In Chaps. 1 and 7 we outlined four possible scenarios for how universities might engage with the SDGs, structured around the two axes of *institutional commitment* (from shallow to deep) and *innovation* (from conventional to bold and ethical). Together they provide a useful heuristic tool for thinking through options for the university and their implications, including what success might look like. In particular, they prompt reflection around two key questions: How deeply will the university commit to the SDGs—now and into the future? How bold and ethical will the innovation culture be—in what areas, why, when and by whom? Only by progressing on both axes will universities be able to achieve the sort of transformative change they need in order to contribute to the transformative change that the world needs.

In Chaps. 4 and 6 we outlined the principles underpinning 'Ethical Innovation' as a normative frame for higher education. These principles are: Responsible, Authentic, Disruptive, Adaptive, Regenerative (RADAR). Regardless of topic area, discipline or institution, research institutions need to become more aware of complexity, uncertainty and the deeply political nature of all research choices and endeavours (including those endeavours that are conspicuous in their absence). This is mirrored in the need for critically reflexive higher education that is *about, for* and *through* the SDGs.

Throughout the book we have argued that understandings and practices in higher education must evolve to better address the need for meaningful real-world change within the context of a rapidly heating and inequitable planet. Universities and society are becoming more complexly entwined, and notions of university success and impact are shifting accordingly. As we have emphasised, the role of the SDGs is two-fold here: representing an agenda to which universities are called upon to contribute, but also a map of the many ways universities themselves need to change. The reciprocal character of the universities and SDGs—intellectually, practically and culturally—means that all universities are implicated in the SDGs as potential 'critical spaces' and agents of change, regardless of their particular characteristics.

Universities: Part of the Problem *and* Solution

Shallow or tokenistic engagement with the SDGs by universities risks legitimating business as usual, thereby perpetuating the processes and systems that are pushing us towards deeper injustice and planetary collapse. Jan Vandemoortele argues that because national governments are likely to—and indeed are beginning to—cherry pick goals and targets to suit themselves and avoid real change, 'civil society, academics, social partners, and other relevant stakeholders must become more involved in target setting, monitoring and critiquing SDG implementation'.[7]

We agree fully with this diagnosis and the call to action for 'academics'. However, it is important not to presume that academics are not as guilty of cynical, inauthentic engagement with the SDGs as any others.

Universities' strongly vested interest in novelty and techno-centric innovation, often individualistic belief in a narrow conception of academic freedom, and uncritical endorsements of research impact, mean that they are often in the thick of unsustainable and unjust business-as-usual activities and visions, such as unending growth in research grant income.

At the same time, universities have a unique capacity to take up Barry's call to 'identify and reduce existing *unsustainabilities*' and to help articulate and achieve a 'future sustainable development path'. As we have discussed in the previous chapters, this poses real challenges for universities and all of us working within them. What is needed in universities is not only more effective means of generating impact, but a more discerning analysis of *what impact is needed* given the impacts that have been generated (intentionally and unintentionally) to date. We also need more robust appreciation of the role of resistance, avoidance and strategic ignorance in the *politics* of unsustainability and injustice.

Such politics does not begin outside of the walls of the university with policy-makers, other 'research end-users' and graduates, who often seem to refuse to understand or adopt our findings or teachings. It is firmly at work within universities, working through myriad channels from research funding to peer review, course offerings to curriculum details, HR choices to procurement decisions, institutional messaging to investment portfolios. It is evident in the long histories of universities in colonial and industrial development, in driving and using the Great Acceleration to their own advantage.

There is growing attention to the many ways in which 'mainstream universities are currently more part of the problem than they are of the solution'. Olivia Bina and Levinia Pereira and others from the EU researcher-practitioner network INTREPID argue that the higher education sector and individual universities are deeply complicit in generating the 'Anthropo-Capitalocene' (a term they use to combine the systems insights of Anthropocene science with the political economy insights of the Capitalocene term, one that locates the drivers for the Anthropocene in capitalism).[8] Fundamental here is the pervasive idealisation of economic growth and its far-reaching effects on knowledge production and education. In terms of research, Bina and Pereira endorse South African scholar Archille Mbembe's assertion that 'university research is complicit in the destruction of the natural world and in the emergence of a new techno-racism'.[9]

Helping enable the use of universities for regressive ends is the evacuation of moral considerations from university decision-making and activities in the name of a purported objectivity and pragmatism. Bina and Pereira argue that:

> By generally omitting (or denying) a space for subjectivity—especially in setting narrowly defined ways of knowing—and related inner change pathways, universities reduce the space to explore the full range of knowing and competencies needed to address the Anthropo-Capitalocene interdependent crises.[10]

Such competencies are frequently absent not only among university graduates, but staff, or at least those in key management roles. Too often questions such as mission, purpose and ethics tend to be reduced to, or dismissed as, mere branding or compliance matters. Universities are at the heart of the knowledge politics that have generated the current crises. The question remains whether they can be at the heart of positive alternatives.

A World in Crisis, Should We Work on Hope?

It is difficult to fully digest—let alone muster up the wisdom and courage—to confront the scope and scale of the challenges the SDG agenda canvasses and those that need to be addressed alongside it. Yet it is also increasingly hard, if not impossible, to ignore that we live in a world in which every one of the crises that the SDGs point to must be addressed. Given this, is it still legitimate to hope for positive outcomes? The many creative responses to these pressures that are emerging around the world suggest to us that it is.

> Today's crises ... present opportunities to move beyond the conventional "solutions" of coping and accommodating, managing and adapting, resisting

and reforming. They create space for social and economic experimentation, new political alliances, new cultural narratives, and alternative social and socio-ecological relations. In short, these crises may give rise to new modes of being in the world that can move us toward a more sustainable and egalitarian future. But how are these new modes of being created and how can activist scholars engage with and support them?[11]

Hope can be understood in different ways. As a verb—*to hope*—the emphasis is on the activity of hoping in the present, whereas the noun *hope* shifts the focus towards the future and what is hoped for. An invitation to think and a provocation to act, hope has been central to social and environmental struggles. Ernst Bloch's *The Principle of Hope* (1950's) discusses utopian hope as the Not-Yet-Consciousness and the multiple principles of a 'utopian homeland' of social justice. In *Pedagogies of Hope* Paulo Freire describes hope as an ontological need. 'The future isn't something hidden in a corner. The future is something we build in the present.'[12] He was writing in the 1970s, but his insight equally applies now. As cultural geographer Lesley Head observes, more than ever, hope needs to be a deliberate practice.[13]

Recognition of the value of hope and utopian imaginaries for social transformation is not new, as highlighted in the previous chapters (e.g. The Good University). While sustainable development remains ambiguous and imperfect, and hopeful sustainable development imaginaries remain on the margins, at base the idea of sustainable development is infused with 'a sense of hope that we can each improve the future well-being of ourselves, each other and the environment'.[14] A growing number of people are helping remake and create new imaginaries of sustainable development through their everyday practices, often engaging in inventive ways with seemingly rigid ideas, politics and realities, as well as forming new and unusual alliances. As Mike Davis argues:

> to raise our imaginations to the challenge of the Anthropocene, we must be able to envision alternative configurations of agents, practices and social relations, and this requires in turn, that we suspend the politico-economic assumptions that chain us to the present.[15]

The SDG agenda is explicitly a Transformation Agenda, one that 'will require deep, structural changes across all sectors in society'.[16] For this reason, and all the discussion, debates, failures, lessons, gains and motivation they have generated already, the SDGs are an important resource and guide for the task of remaking sustainable development. So too are universities. As institutions with the privilege of access to knowledge, ideas, networks and dialogue, as well as often unusual degrees of autonomy, universities can and need to contest the 'dictatorship of no alternatives'.[17] As *education* institutions, they can offer alternatives and teach hope to students. In the words of Paul Warwick and colleagues, 'within troubled times of global challenge, hope is an imperative within education'. As they argue, we need to repurpose higher education 'to empower students with the hope of a positive anticipation that more sustainable futures are possible'.[18]

As *research* institutions, universities have an unusually powerful role in shaping the future. Every university has an opportunity to give 'analytical time and space to counter-normative practices' and help open up 'possibilities of alternative futures' if they so choose.[19] To do so, they need to loosen their grip on entrenched assumptions and ways of doing things and shake the habit of 'a paranoid critical stance' that casts anything else—notably anything more hopeful—as 'naive, pious or complaisant'.[20]

Rather than being rooted in dogma, universities can more overtly offer spaces in which ambiguity and ambivalence are acknowledged, and reparative practices of knowing are pursued. As discussed in Chap. 4 on ethical innovation, this means critically reflecting on the way in which our knowledge production practices are, or are not, *(re)generative* of better futures and attending to the atmosphere (both in terms of the Earth's air and society's moods and ambitions) that we are inevitably helping create. Pollution, despair and cynicism—or oxygen, hope and resolve?

An atmosphere thick with cynicism is debilitating. Instead, as Paulo Freire put it, 'We need critical hope the way a fish needs unpolluted water'.[21] To aim for and practice hope is not to imagine it is sufficient. As Freire continues, critical hope 'is necessary but it is not enough. Alone it does not win.'[22] Nor is a commitment to hope simply an effort to wish away the difficulties of the world, deny ironies or 'sidestep the messy world of practice'.²³ It is to face such difficulties and mess with compassion and commitment. It is to appreciate that the state of the world and universities' role is 'an open-ended story' that we are helping tell through what we choose to think, say and do.²⁴

Other more sustainable development futures are still possible. In facing the openness of the future, universities need to face important questions of the sort passionately articulated by Boaventura de Sousa Santos:

- Modern universities have been a product and a producer of specific models of development, including training elites and providing knowledge and ideology. Can the university contribute to dialogues of different models of development and refound its mission?
- Can the university acknowledge that knowledge is everywhere, not just behind its walls?
- In particular, can it recognise that human understanding of the world far exceeds the Western ways of thinking that dominate the structure and content of global higher education?

The work of the Community Economies Collective²⁵ and their related research networks, for example, demonstrate that other, more just and ecologically sustainable, worlds are possible. This involves 'everyday people in everyday practices' taking part in re-thinking and re-enacting economies: to re-imagine an economic politics that allows us to think creatively to make new economies, building on the alternative economic practices that already exist in the shadow of the capitalist Economy all over the world.²⁶ Notably, this Collective is a collaboration between universities and local communities across diverse parts of the globe and demonstrates the sort of relational ethics that is needed.

In their manifesto *Take Back the Economy*, some of the founders of the Community Economies Collective, J.K. Gibson-Graham, Jenny Cameron and Stephen Healy, underline the importance of hope in their work, illustrating how it helps connect their twin focus on the very big and the very small, on the very ambitious and the very practical. Some of their recent work includes co-developing progressive and useful impact indicators with communities, contributing to the work of the UN Inter-Agency Task Force on Social and Solidarity Economy to embed the social and solidarity economy into the SDGs in belated recognition of its neglect in the original formulation of the SDGs.[27] This whole realm of activity demonstrates the potential for academics to work across scales in creative and experimental ways that draw on and feedback on the SDGs to help co-create more positive futures. It also demonstrates the way in which some academics are already working from within universities to help generate dialogue about different models of development, in the way de Sousa Santos notes is needed.

Avoiding Traps

The SDGs agenda can help universities take the action that is urgently needed by encouraging them to avoid the two traps that many of them cohabit or flip between. The first trap is being disengaged from the 'real world'; what Kamola associated with 'a global imaginary' that views the Earth from space. Here, the SDG agenda—while at first blush part of the global imaginary because of its international reach—actually challenges the notion that any of us are divorced from the planet or able to pronounce upon the world from afar. In contrast to the assumption that 'development' is just something for poor countries, it enrols all nations and all organisations in sustainable development and requires universities to look inward as well as outward.

Those of us within universities need to call out dismissive or shallow engagement with the SDGs, particularly that which presumes that the aim of such engagement is to benevolently assist 'those people over there'. We need to demonstrate and advocate for more transformational approaches that begin by identifying universities' role at the centre of the problem and change them from within. As Maori Hirini Matunga powerfully highlights, far from being transformative, tokenistic engagement instead becomes:

> An alienated and alienating blah, that, rooted 'deep down' in its colonial past and present—actually knows the problem, but in a form of soporific amnesia has airbrushed it out of existence, because confronting it requires facing up to its own history, its own complicity with the colonial project, and its ongoing marginalisation and dispossession of the very communities

it actually needs to engage. … Is it even trying to 'call out' power for what it is? Or has it become so deprived of its dimensions of justice and emancipatory action that it has become a functionary of the economic, political and often racial elite, in what remains an obstinately colonial, settler dominant, market-driven system?

The second trap that universities fall into is that (in an attempt to dodge criticisms of being self-indulgent 'ivory towers') many have strenuously worked to demonstrate their relevance to the real world—but *mistaken what that world is*. While some universities are usefully reviving lost, centuries-old and largely non-economic notions of what universities are and for, many have interpreted relevance in terms of the dominant contemporary discourse that equates the capitalist market with reality. Thus, attempts at 'engagement' and 'impact' are overly oriented towards technological solutions and generating financial returns on investment.

By framing universities and their research partners and graduate employers in economic, hyper-modernist (and often nationalistic) terms, this reduction of higher education to a capitalist activity disguises and justifies the negative effects it is having in the world (e.g. supporting processes that are materially intensive and discriminatory), and marginalises higher education's far broader public value. More generally, this misreading of higher education potential perpetuates the dominant economic discourse that has appropriated and perverted the very notion of value, and perverted the role of government and other institutions such as universities by defining 'value creation' in terms of rapid, content-neutral economic gains.

As economist Mariana Mazzucato argues in *The Value of Everything*, public institutions (including universities) need to 'reclaim their rightful role as servants of the common good' by challenging the logics and metrics that orient them to the short term and underplay their capacity to proactively germinate, nurture and shape markets, not just respond to them.[28] She concludes that a 'new economics: an economics of hope' needs to begin with the fact that 'the creation of value is collective' and then develop 'a dynamic division of labour focused on the problems that twenty-first-century societies are facing'.[29] Universities, she underlines, are crucial to this effort.

The SDG agenda helps universities avoid the self-defeating trap of reading the world and their own role in it through a narrow capitalist lens. It draws universities out of their myopic focus on themselves and their coterie of current industry partners and graduate employers to look further afield to the troubled world and futures they are inadvertently helping create. It begins to unsettle the notions that the economy exists as an independent entity disconnected from the social or environmental, and that value can be divorced from what an activity actually does in the world. Mazzucato advocates strongly for the SDG agenda as a *mission* around which institutions and other actors should coordinate.[30]

The SDG agenda offers a response to the fact that 'to offer real change we must go beyond fixing isolated problems' and instead develop a framework that allows us to collectively and effectively 'work for the common good'.[31] For universities, Patsy Healy suggests, this is about using current instabilities and crises 'in a strategic way, as an opportunity to take stock, to re-think policies, projects and practices, and to build the intelligence and coalitions which could bring future benefits for the many not just the few in our localities'.[32]

Another Future Is Possible

Universities are animated by an inherent future focus, one that is core to their developmentality. The horrors, risks and uncertainties of the Anthropocene do nothing to dim this focus on the future; indeed they underline the need to take the future more seriously than ever. But they do blur our vision and scramble our taken-for-granted maps. They wake us up to the fact that in chasing growth without care for direction, we have already lost our way. In this way, the Anthropocene also demands that we look backwards, and into our institutions and selves, to understand the situation we are in and ask what it is we are trying to develop.

Thinking more carefully about 'the future' is one of the core directives of the SDG agenda. As we do so, we draw on some of the useful knowledge and tools we already have at our disposal, bucking against the trend for universities to manage themselves without ever using the expertise they house to help address their own problems. Of particular use is not

8 Sustainable Futures

only the work of highly engaged academics such as Mariana Mazzucato, Patsy Healy or the many others we refer to in this book, but also the 'futures thinking tools' developed over the last few decades—noting that the tools themselves are agnostic to what futures are envisaged and created, and so need to be accompanied by careful analysis of directionality and impact.

A simple but compelling approach is offered by the Three Horizons model of Bill Sharpe, now used widely by the International Futures Federation. Its adoption by another highly engaged academic—renegade Oxford University economist Kate Raworth, author of *Doughnut Economics* and advocate for creating more just and regenerative economies—demonstrates how valuable it is in trying to envisage pathways towards more progressive futures. The Three Horizons foresight model[33] proposes that we can imagine elements or seeds of different futures existing in the present. These different 'worlds' are summarised in the model as three horizons (see Fig. 8.1). Horizon 1 is Business as Usual, and when viewed from the present, it is often all that we can see or even imagine. Characterised by 'sustaining' (not necessarily sustainable) innovations, it is focused on sustaining Business as Usual and is poorly adapted to

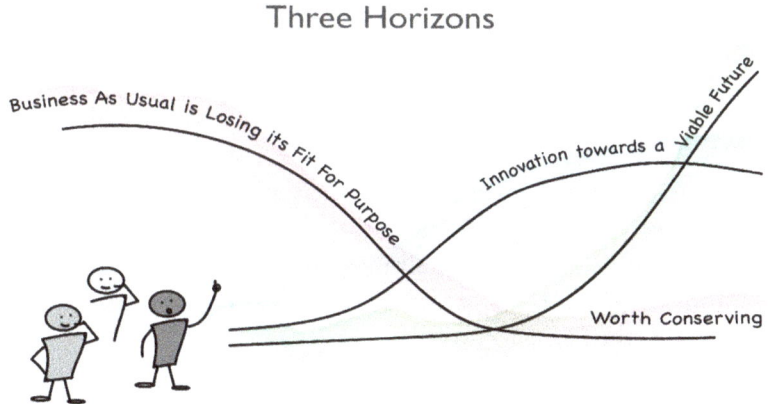

Fig. 8.1 The Three Horizons framework. (Adaptation by the social enterprise, The H3 Uni https://www.h3uni.org/practices/foresight-three-horizons/)

emerging conditions. Not long into the future, it falls away to a greater or lesser degree.

Horizon 2 is about emerging positive changes, seeds of which are evident in the present and quickly grow, but do not, without further help, drive systemic change. Horizon 3 is the more fully transformed world we want to cultivate. Generated through a strategic combination of innovations, structural shifts and dismantling of barriers, it represents foundational change and great upheaval at first. Because it is far better adapted to contemporary and emerging challenges, though, ultimately it is the more sustainable in the long term.

Arguably the SDG agenda is a Horizon 2 intervention—disruptive but not in itself (as a mere agenda or plan) transformational. The question then is whether its (non)implementation will allow it to be captured by the currently dominant Horizon 1, or whether we will be able to harness it to H3 and turn into a H2+ stepping stone to long term positive transformation. Experiences to date with colonial, international and sustainable development, plus evidence of much existing engagement with the SDG agenda, suggest that we cannot underestimate the risk of it being co-opted and becoming what Sharpe and colleages call a H2- pathway, one that looked promising but ultimately becomes entweined with and declines with Horizon 1. But as we have argued in this book, the SDG agenda itself does not predetermine how it is interpreted and implemented. For those of us in universities at least, it *offers* a pathway to much-needed positive change; the question is whether we use it.

So, what does a future, Horizon 3 type university look like and how can the SDGs help? Olivia Bina, Levinia Pereira and the INTREPID network, mentioned above, have examined this question of a Horizon 3 type university in a hopeful but critical register. They offer a vision of future universities as places with six interrelated characteristics (Fig. 8.2).[34] We outline them here, elaborating on their vision by underlining the way it aligns with the SDG agenda:

1. *A place of 'maximum leverage'*: Universities are places in which Donella Meadow's most powerful leverage points for systems change—reassessing goals, reassessing paradigms and worldviews and appreciating the value of different worldviews—are discussed, strengthened and practiced. As Bina and Pereira put it, 'we imagine universities as

8 Sustainable Futures   261

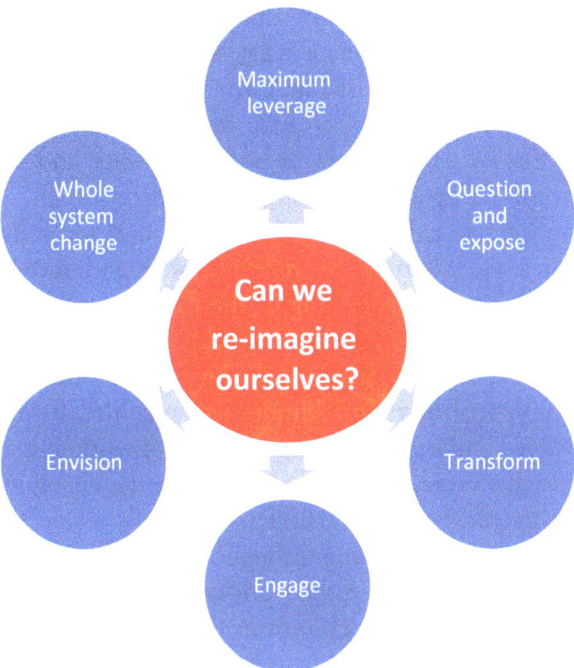

Fig. 8.2 Re-imagining the future of the university—six priorities. (Adapted from Bina and Pereira 2020, p. 22)

places where the uncomfortable problems and unorthodox solutions, such as beyond-GDP (gross domestic product) are explored'.[35] Such a role for universities is vital to their capacity to help drive transformational change for and beyond the SDG agenda. The value of the SDGs is they help redress the implicit goals driving dominant development agendas, including in higher education. While universities are heavily invested in historical trajectories, they can become places where, in the spirit of criticality, entrenched worldviews are critiqued, and their positive elements are renovated and combined with vital elements of alternative perspectives.

2. *A place to question and expose*: Universities are places that foster and demonstrate critical thinking, questioning biases and assumptions, exposing implicit goals and intentions, and ultimately confronting the direct and indirect drivers of the Anthropo-Capitalocene in order to

'phase out supporting socially and ecologically unsustainable systems'.[36] This is at the heart of what is needed both to advance the SDG agenda and to call out its own weaknesses. Detailed analysis, sophisticated dialogue and creative experimentation are needed to untangle the factors involved in unsustainability and injustice—all tasks that call for greater university involvement.

3. *A place to transform*: Universities can help transition individuals and society to a more self-aware, reflexive, wise and sustainable basis, including a deep understanding of the fundamental interdependencies of humans and the rest of the world. Bina and Pereira note that the field of Ecologically Sustainable Development that SDG 4 advocates for is crucial here. In addition, we argue that helping question what is valued and demonstrating the value of wisdom is another crucial way higher education can help generate the enabling conditions for achieving and exceeding the SDGs.

4. *A place to engage*: As discussed throughout this book, universities need to re-think their position in the world and in particular challenge the outdated imaginary in which universities are separate to society and the planet. Many are beginning to do so, and it is increasingly apparent that universities *can* help foster new ways of engaging with broader society, including co-production of knowledge and collaboration with local communities. Reshaping themselves as places for diverse groups to engage on shared problems and to pursue the common good is central to how universities can help progress the SDG agenda.

5. *A place to envision*: Universities offer a space in which diverse groups can come together to envisage and create more sustainable, just futures. This includes engaging with the SDG 'transforming the world' agenda, both to help turn the SDG vision into a reality and to push it further, using it as a Horizon 2 stepping-stone towards a truly transformational Horizon 3 world. Working in this way requires university members of all sorts to take seriously their role to inspire as well to inform, in keeping with a Freirean pedagogy of hope: 'a mode of hoping … in the possibility of attaining the goal we dream up [that] lies … in the inspirational qualities of the goal itself, in its capacity to … expand the horizons of possibility'.[37]

6. *A place of whole-of-system change*: To take on the SDG agenda, universities need to change themselves. This is about appreciating the far-reaching physical and social effects that universities generate every minute of the day at multiple scales. If universities are to become less of the problem and more of the solution, they need to not only help others, but change inside out. In addition to weaving SDGs through their curricula or running SDG hackathons or badging research projects with specific SDGs, this requires altering their 'physical, organizational and institutional structures' and 'overall governance and management practices' to ensure that they are working for environmental sustainability and social justice in all they do.[38]

SDGs: A Witness Statement for Higher Education

Like many people around the world, we two Australian authors have recently emerged from catastrophic bushfires, floods, heatwaves and drought. We are living *in* climate change. For all of us, climate change is not some distant agenda 'out there', it's here and now.[39] Combined with the ongoing impacts of COVID-19, including the worsened social and economic inequalities that are in turn deepening climate change vulnerabilities, the need for deep social change is more apparent than ever. One of the reasons we advocate for the SDGs is because they explicitly address the need to take urgent action on climate change and call for the transformative change required to reduce both greenhouse gas emissions and vulnerabilities in order to sustain life on the planet. On this and other issues, the SDGs are the world's witness statement to the planetary and social condition—drawing attention to what needs to be attended to at both local and global scales.

We ignore them at our peril.

Systems change of the sort the SDGs demand and universities require is no quick fix. But quick work is needed to commence it now. For self-serving reasons alone, universities need to rapidly begin transforming.

They are already facing questions from potential students about the value of university degrees in equipping them for the future and are already under pressure to better demonstrate their positive impact. The time is now to renew their purpose and revitalise their role in society. And one such role has to be helping to scale up the SDG agenda from a niche or abstract concept into the culture, literacy and workings of institutions, including but not limited to higher education.

The challenge of changing universities is not to be underestimated. As we argued in Chap. 3, they are highly resilient institutions. Some are likely to be deeply committed to change, but still not do much differently, other than reshape a few processes, leaving untouched key areas such as leadership and business decisions. As critical education scholars have long pointed out, formal education is a mechanism of social reproduction, and so while degrees are sold to individuals as a route to social mobility not social change, universities risk reinforcing existing hierarchies, structures and problems, as well as the social anxieties and ambitions that legitimise them. The neoliberal university's emphasis on changing product and customer specifications (e.g. through course marketing and/or Net Promoter Scores) must shift towards changing university systems, goals and paradigms—including the culture itself—so that society's needs, including planetary integrity, are more effectively met. This 'third generation' approach to impact is not just about new content but new structures, processes and ethos, including the need to:

- *Redirect* the potential role and contribution of universities in addressing and reducing global socioeconomic and environmental inequalities as the central priority.
- *Shift* the focus as a sector from competition to collaboration through partnerships and networks across disciplinary areas and diverse stakeholders, acknowledging that, as Audre Lorde has argued, the transformative challenge of the SDGs is to define and empower not to conquer and divide.
- *Work* across boundaries to link up and scale up efforts across different issues, identify synergies and tensions and foster a new way of working.

- *Balance* the quest for new income sources and resources and the need to do more with less, with increasing public commitment and belief in the role of the university.

This is the transformative SDG agenda we imagine, animated by the critical, regenerative politics needed to reshape the dominant unsustainable development trajectories in higher education and society more broadly. More sustainable worlds are still possible, and higher education has a vital role to play.

Notes

1. Marginson, S (2010) Marginson, S. (2010), 'The University: Punctuated by Paradox', *Academic Matters*, May, p. 14–18.
2. Barry, J. (2013). *The Politics of Actually Existing Unsustainability: Human flourishing in a climate-changed, carbon-constrained world*. Oxford, Oxford University Press. p. 6, 14.
3. Aristotle—cited in Lewis and Maslin (2019).
4. See for example: Forestier, O., Kim, R.E. (2020) Cherry-picking the Sustainable Development Goals: Goal prioritization by national governments and implications for global governance. *Sustainable Development* 28, p. 1269–1278. Siegel, K.M., Lima, M.G.B. (2020) When international sustainability frameworks encounter domestic politics: The sustainable development goals and agri-food governance in South America. *World Development* 135, p. 105053.
5. Sonetti, G, Brown, M and Naboni, E (2018) About the Triggering of UN Sustainable Development Goals and Regenerative Sustainability in Higher Education, *Sustainability*, 11(254), p. 1–17.
6. SDSN (2018) Getting Started with the SDGs.
7. Vandemoortele, J. (2018) From simple-minded MDGs to muddle-headed SDGs. *Development Studies Research* 5, p. 83–89.
8. Bina, O. and L. Pereira (2020). Transforming the Role of Universities: From Being Part of the Problem to Becoming Part of the Solution. *Environment: Science and Policy for Sustainable Development* 62(4): p. 16–29.

9. Mbembe, A. (2016) The age of humanism is ended. *Mail and Globe.* https://mg.co.za/article/2016-12-22-00-the-age-of-humanism-is-ending/
10. Bina and Pereira (2020), p. 22.
11. Burke, B., and Boone W. Shear. 2014. Introduction: Engaged Scholarship for Non-Capitalist Political Ecologies. *Journal of Political Ecology* 21: p. 127–44.
12. Freire, P. (2016). *Pedagogy of Hope: Reliving Pedagogy of the Oppressed.* London: Bloomsbury.
13. Head, L. (2019) *Hope and Grief in the Anthropocene: Re-conceptualising human-nature relations.* Routledge, London.
14. Ibid., p. 116.
15. Mike Davis (2010, p. 45).
16. Sachs, J. D., G. Schmidt-Traub, M. Mazzucato, D. Messner, N. Nakicenovic and J. Rockström (2019). "Six transformations to achieve the sustainable development goals." *Nature Sustainability* 2(9): 805–814. p. 805.
17. Unger, R.M. (2005). *What should the Left propose?* London: Verso.
18. Warwick, P., et al. (2019). The Pursuit of Compassionate Hope: Repurposing the University Through the Sustainable Development Goals Agenda. *Higher Education and Hope*, Springer: p. 113–134. p. 114.
19. Roseneil, S. (2011). Criticality, Not Paranoia: A Generative Register for Feminist Social Research. *NORA—Nordic Journal of Feminist and Gender Research* 19(2): 124–131. p. 130.
20. Sedgwick, Eve Kosofsky (2003) *Paranoid Reading and Reparative Reading, or, You're So Paranoid, You Probably Think This Essay Is About You, in Touching Feeling: Affect, Pedagogy, Performativity.* Duke University Press: Durham & London. p. 126.
21. Freire, P. (2016). *Pedagogy of Hope: Reliving Pedagogy of the Oppressed.* London: Bloomsbury. p. 2.
22. op cit.
23. Spicer, A. (2018). *Business Bullshit.* New York: Routledge. p. 107.
24. Hyvönen, A.-E. (2019). Pedagogies of Hopefulness and Thoughtfulness: The Social-Political Role of Higher Education in Contemporary Societies. *Higher Education and Hope,* Springer: p. 21–48.
25. See https://www.communityeconomies.org
26. Ibid.
27. See https://unsse.org/sse-and-the-sdgs/

28. Mazzucato, M. (2018) *The Value of Everything: Making and taking in the global economy.* Penguin, London. p. 266.
29. Ibid., p. 280.
30. Mazzucato, M. (2020) *Mission Economy: A moonshot guide to changing capitalism.* Allen Lane, London.
31. Mazzucato, The Value of Everything, p. 271.
32. Patsy Healey (2010, p. 15).
33. Not to be confused with the 'growth model' of the same name used in the corporate world.
34. Bina and Pereira (2020), p. 22.
35. Ibid., p. 23.
36. Op cit.
37. Webb, D. (2013). Pedagogies of hope. *Studies in Philosophy and Education* 32(4): p. 397–414.
38. Stephens, J. C. and A. C. Graham (2010). "Toward an empirical research agenda for sustainability in higher education: exploring the transition management framework." *Journal of cleaner production* 18(7): p. 615.
39. Steele, W (2020), *Planning Wild Cities: Human-Nature Relationships in the Urban Age*, New York, Routledge.

Bibliography

Agbedahin, A. V. (2019). Sustainable Development, Education for Sustainable Development, and the 2030 Agenda for Sustainable Development: Emergence, Efficacy, Eminence, and Future. *Sustainable Development, 27*(4), 669–680.
Allenby, B., & Sarewitz, D. (2011). *The Techno-Human Condition*. MIT Press.
Angus, I. (2016). *Facing the Anthropocene: Fossil Capitalism and the Crisis of the Earth System*. NYU Press.
Archer, C., & Willi, A. (2012). *Opportunity Costs: Military Spending and the UN's Development Agenda. A View from the International Peace Bureau*. International Peace Bureau.
Barnett, C. (2018, April 11). The Financialisation of Higher Education and the USS Pension Dispute. *Medium*. https://medium.com/ussbriefs/the-financialisation-of-higher-education-and-the-uss-dispute-9231b9458699
Barnett, C., & Parnell, S. (2016). Ideas, Implementation and Indicators: Epistemologies of the Post-2015 Urban Agenda. *Environment & Urbanization, 28*(1), 1–13.
Barry, A., Born, G., & Weszkalnys, G. (2008). Logics of Interdisciplinarity. *Economy and Society, 37*, 20–49.

Barry, J. (2013). *The Politics of Actually Existing Unsustainability: Human Flourishing in a Climate-Changed, Carbon-Constrained World.* Oxford University Press.

Bart, M., Michelsen, G., Rieckmann, M., & Thomas, I. (2016). *Routledge Handbook of Higher Education for Sustainable Development.* Routledge.

Barth, M., & Burandt, S. (2013). Adding the "e-" to Learning for Sustainable Development. *Sustainability, 5,* 2609–2622.

Bayne, S., Evans, P., Ewins, R., Knox, J., Lamb, J., McLeod, H., O'Shea, C., Ross, J., Sheail, P., & Sinclair, C. (2020). *The Manifesto for Teaching Online.* MIT Press.

Beck, U., Giddens, A., & Lash, S. (1994). *Reflexive Modernisation.* Polity.

Bell, S., Douce, C., Caeiro, S., Teixeira, A., Martín-Aranda, R., & Otto, D. (2017). Sustainability and Distance Learning: A Diverse European Experience? *Open Learning: The Journal of Open, Distance and e-Learning, 32*(2), 95–102.

Ben-Arieh, A., & Goerge, R. (2006). *Indicators of Children's Wellbeing: Understanding Their Role, Usage and Policy Influence.* Springer.

Bina, O., & Pereira, L. (2020). Transforming the Role of Universities: From Being Part of the Problem to Becoming Part of the Solution. *Environment: Science and Policy for Sustainable Development, 62*(4), 16–29.

Blades, G. (2019). *Walking Practices with/in Nature(s) as Ecopedagogy in Outdoor Environmental Education: An Autophenomenographic Study,* Unpublished PhD Thesis, School of Education, College of Arts, Social Sciences and Commerce, La Trobe University, Victoria, Australia.

Blühdorn, I. (2011). The Politics of Unsustainability: COP15, Post-ecologism, and the Ecological Paradox. *Organization & Environment, 24,* 34–53.

Blythe, J., Silver, J., Evans, L., Armitage, D., Bennett, N., Moore, M., Morrison, T., & Brown, K. (2018). The Dark Side of Transformation: Latent Risks in Contemporary Sustainability Discourse. *Antipode, 50,* 1206–1223.

Boldeman, L. (2007). *The Cult of the Market: Economic Fundamentalism and Its Discontents.* ANU Press.

Bonneuil, C., & Fressoz, J. P. (2015). *The Shock of the Anthropocene: The Earth, History and Us.* Verso.

Braidotti, R. (2013). *The Posthuman.* Polity Press.

Brown, E., & McCowan, T. (2018). Buen Vivir: Reimagining Education and Shifting Paradigms. *Compare: A Journal of Comparative and International Education, 48*(2), 317–323.

Brugmann, R., Cote, N., Postma, N., Shaw, E. A., Pal, D., & Robinson, J. B. (2019). Expanding Student Engagement in Sustainability: Using SDG- and CEL-Focused Inventories to Transform Curriculum at the University of Toronto. *Sustainability, 11*(2), 530.
Brundtland Commission. (1987). *Our Common Future.* Oxford University Press.
Buhmann, K., Jonsson, J., & Fisker, M. (2019). Do No Harm and Do More Good Too: Connecting the SDGs with Business and Human Rights and Political CSR Theory. *Corporate Governance: The International Journal of Business in Society, 19*(3), 389–403.
Buil-Fabrega, M., Casanovas, M. M., Ruiz-Munzon, N., & Leal, W. (2019). Flipped Classroom as an Active Learning Methodology in Sustainable Development Curricula. *Sustainability, 11*(17), 4577.
Burke, B., & Shear, B. W. (2014). Introduction: Engaged Scholarship for Non-capitalist Political Ecologies. *Journal of Political Ecology, 21,* 127–144.
Bush, V. (1945). *Science—The Endless Frontier.* A Report to the President by Vannevar Bush. Director of the Office of Scientific Research and Development, United States Government Printing Office, Washington, DC.
Calisto Friant, M., & Langmore, J. (2015). The Buen Vivir: A Policy to Survive the Anthropocene? *Global Policy, 6*(1), 64–71.
Campbell, S. (1996). Green Cities, Growing Cities, Just Cities? Urban Planning and the Contradictions of Sustainable Development. *Journal of the American Planning Association, 62*(3), 296–312.
Carrol, N. (2002). Why Horror? In N. Jancovich (Ed.), *Horror: The Film Reader.* Routledge.
Cash, D. W., Clark, W. C., Alcock, F., Dickson, N. M., Eckley, N., Guston, D. H., Jäger, J., & Mitchell, R. B. (2003). Knowledge Systems for Sustainable Development. *Proceedings of the National Academy of Sciences of the United States of America, 100*(14), 8086–8091.
Chandler, D., Grove, K., & Wakefield, S. (2020). *Resilience in the Anthropocene: Governance and Politics at the End of the World.* Taylor and Francis.
CMU. (2020). *Voluntary University Review of the Sustainable Development Goals.* Pittsburgh: Carnegie Mellon University Sustainability Initiative. Accessed Online at https://www.cmu.edu/leadership/the-provost/provost-initiatives/cmu-vur-2020
Coffey, B. (2016). Unpacking the Politics of Natural Capital and Economic Metaphors in Environmental Policy Discourse. *Environmental Politics, 25,* 203–222.

Collard, R.-C., Dempsey, J., & Sundberg, J. (2015). A Manifesto for Abundant Futures. *Annals of the AAG, 105*, 323.

Connell, R. (2019). *The Good University: What Universities Actually Do and Why It's Time for Radical Change.* University of Monash Press.

Coulton, C., & Korbin, J. (2007). Indicators of Child Wellbeing Through a Neighborhood Lens. *Social Indicators Research, 84*(3), 3.

Craig, R., Amernic, J., & Tourish, D. (2014). Perverse Audit Culture and Accountability of the Modern Public University. *Financial Accountability & Management, 30*, 1–24.

Croissant, J. L. (2015). Routine, Scale, and Inequality: Introduction to the Special Issue on Ethics, Organizations, and Science. *Science, Technology, & Human Values, 40*, 167–175.

de Beauvoir, S. (2015). *The Second Sex.* Penguin.

De Saille, S. (2015). Innovating Innovation Policy: The Emergence of 'Responsible Research and Innovation'. *Journal of Responsible Innovation, 2*, 152–168.

Death, C., & Gabay, C. (2015). Doing Biopolitics Differently? Radical Potential in the Post-2015 MDG and SDG Debates. *Globalizations, 12*, 597–612.

Deininger, N., Lu, Y., Griess, J., & Santamaria, R. (2019). *Cities Taking the Lead on the Sustainable Development Goals.* Brookings Institution.

Deleuze, G., & Guattari, F. (1987). *A Thousand Plateaus: Capitalism and Schizophrenia.* Athlone Press.

Demssie, Y. N., Wesselink, R., Biemans, H. J. A., & Mulder, M. (2019). Think Outside the European Box: Identifying Sustainability Competencies for a Base of the Pyramid Context. *Journal of Cleaner Production, 221*, 828–838.

Eaton, C., Habinek, J., Goldstein, A., Dioun, C., Santibáñez Godoy, D. G., & Osley-Thomas, R. (2016). The Financialization of US Higher Education. *Socio-Economic Review, 14*, 507–535.

Eaton, C., & Stevens, M. L. (2020). Universities as Peculiar Organizations. *Sociology Compass, 14*, 12768.

Edwards, P. N. (2010). *A Vast Machine: Computer Models, Climate Data, and the Politics of Global Warming.* MIT Press.

Felman, S. (1992). The Return of the Voice: Claude Lanzmann's Shoah. In S. Felman & D. Laub (Eds.), *Testimony: Crises of Witnessing in Literature, Psychoanalysis and History.* Routledge.

Fien, J. (2002). Advancing Sustainability in Higher Education: Issues and Opportunities for Research. *Higher Education Policy, 15*, 143–152.

Filho, W. (Ed.). (2018). *Implementing Sustainability in the Curriculum of Universities: Approaches, Methods and Projects*. Springer Nature.

Fischer, J., & Riechers, M. (2019). A Leverage Points Perspective on Sustainability. *People and Nature, 1*, 115–120.

Forestier, O., & Kim, R. E. (2020). Cherry-Picking the Sustainable Development Goals: Goal Prioritization by National Governments and Implications for Global Governance. *Sustainable Development, 28*, 1269–1278.

Foucault, M. (1977). *Discipline and Punishment: The Birth of the Prison*. Pantheon Books.

Fournier, M. (2014). Lines of Flight. *Transgender Studies Quarterly, 1*(1–2), 121–122.

Freire, P. (1992). *Pedagogy of the Oppressed*. Penguin.

Freire, P. (2016). *Pedagogy of Hope: Reliving Pedagogy of the Oppressed*. Bloomsbury.

Fry, T. (2014). *Cities for a Future Climate*. Routledge.

Fukuda-Parr, S., Yamin, A. E., & Greenstein, J. (2014). The Power of Numbers: A Critical Review of Millennium Development Goal Targets for Human Development and Human Rights. *Journal of Human Development and Capabilities, 15*(2–3), 105–117.

Fuller, S. (2013). Climate Justice and Global Cities: Mapping the Emerging Discourses. *Global Environmental Change, 23*, 914–925.

Funtowicz, S., & Ravetz, J. K. (2018). Post-normal Science. In *Companion to Environmental Studies* (pp. 443–447). Routledge in Association with GSE Research.

Gallagher, S. (2018). Development Education on a Massive Scale: Evaluations and Reflections on a Massive Open Online Course on Sustainable Development. *Policy & Practice: A Development Education Review, 26*, 122–140.

Gandy, M. (2005). Cyborg Urbanization: Complexity and Monstrosity in the Contemporary City. *International Journal of Urban and Regional Research, 29*(1), 26–49.

Gemenne, F. (2015). The Anthropocene and Its Victims. In C. Hamilton, C. Bonneuil, & F. Gemenne (Eds.), *The Anthropocene and the Global Environmental Crisis: Rethinking Modernity in a New Epoch* (pp. 168–174). Routledge.

Gibbs, R. W. (2002). Irony in the Wake of Tragedy. *Metaphor and Symbol, 17*, 145–153, 152.

Gibson-Graham, J. K. (2006). *A Post-capitalist Politics.* University of Minnesota Press.
Gibson-Graham, J. K., & Roelvink, G. (2010). An Economic Ethics for the Anthropocene. *Antipode, 41*(s1), 320–346, 324.
Goddard, J. (2018). The Civic University and the City. In *Geographies of the University* (pp. 355–373). Springer.
Godin, B. (2009). National Innovation System: The System Approach in Historical Perspective. *Science, Technology, & Human Values, 34,* 476–501.
Godin, B. (2015). *Innovation Contested: The Idea of Innovation Over the Centuries.* Routledge.
Goodwin, B., & Taylor, K. (2009). *The Politics of Utopia: A Study in Theory and Practice.* Peter Lang.
Graham, S., & Marvin, S. (2001). *Splintering Urbanism: Networked Infrastructures, Technological Mobilities and the Urban Condition.* Routledge.
Grant, J. (1999). *A Handbook of Economic Indicators.* University of Toronto Press.
Gudynas, E. (2011). Buen Vivir: Today's Tomorrow. *Development, 54*(4), 441–447.
Habermas, J. (1973). *Theory and Practice.* Beacon Press.
Hajer, M., Nilsson, M., Raworth, K., Bakker, P., Berkhout, F., de Boer, Y., Rockström, J., Ludwig, K., & Kok, M. (2015). Beyond Cockpit-ism: Four Insights to Enhance the Transformative Potential of the Sustainable Development Goals. *Sustainability, 7*(2), 1651–1660.
Hák, T., Janoušková, S., & Moldan, B. (2016). Sustainable Development Goals: A Need for Relevant Indicators. *Ecological Indicators, 60,* 565–573.
Halovitch, H. (2019). Vampires and Ratko Mladić: Balkan Monsters and the Monstering of People. In J. Lee, H. Halilovitch, A. Landau-Ward, P. Phipps, & R. Sutcliffe (Eds.), *Monsters of Modernity: Global Icons for Our Critical Condition.* Kismet Press.
Haraway, D. (2016). *Staying with the Trouble: Making Kin in the Chthulucene.* Duke University Press.
Hart, P. (1993). *Research in Environmental Education: Engaging the Debate.* Deakin University.
Head, L. (2019). *Hope and Grief in the Anthropocene: Re-conceptualising Human-Nature Relations.* Routledge.
Healey, P. (2010). Re-thinking the Relations Between Planning, State and Market in Unstable Times. In P. Paolo & O. Vitor (Eds.), *Planning in Times of Uncertainty.* FEUP Edicoes.

Hegarty, K., & Holdsworth, S. (2016). Towards a Scholarship of Curriculum Change: From Isolated Innovation to Transformation. In M. Barth, G. Michelsen, M. Rieckmann, & I. Thomas (Eds.), *Routledge Handbook of Higher Education for Sustainable Development*. Routledge.
Hellström, T., & Jacob, M. (2005). Taming Unruly Science and Saving National Competitiveness: Discourses on Science by Sweden's Strategic Research Bodies. *Science, Technology, & Human Values, 30*, 443–467.
Hess, C., & Ostrom, E. (Eds.). (2007). *Understanding Knowledge as a Commons*. MIT Press; Federici, S. (2009). Education and the Enclosure of Knowledge in the Global University. *ACME, 8*(3), 454–461.
Hillier, J. (2007). *Stretching Beyond the Horizon: A Multiplanar Theory of Spatial Planning and Governance*. Ashgate.
Hillier, J. (2011). Strategic Navigation Across Multiple Planes: Towards a Deleuzean-Inspired Methodology for Strategic Spatial Planning. *Town Planning Review, 82*, 503–527.
Hoernig, H., & Seasons, M. (2004). Monitoring of Indicators in Local and Regional Planning Practice: Concepts and Issues. *Planning Practice and Research, 19*(1), 84.
Hope, J. (2021). The Anti-politics of Sustainable Development: Environmental Critique from Assemblage Thinking in Bolivia. *Transactions of the Institute of British Geographers*. Online First.
Hyvönen, A.-E. (2019). Pedagogies of Hopefulness and Thoughtfulness: The Social-Political Role of Higher Education in Contemporary Societies. In *Higher Education and Hope* (pp. 21–48). Springer.
Ibrahim, Z. (2020, January 13). Universities Divesting from Fossil Fuels Have Made History But the Fight Isn't Over. *The Guardian*. Accessed Online at https://www.theguardian.com/education/2020/jan/13/universities-divesting-from-fossil-fuels-have-made-history-but-the-fight-isnt-over
IFLA. (2020). The 6 P's of Libraries and the SDGs. Accessed on https://blogs.ifla.org/lpa/2020/10/12/the-6-ps-of-libraries-and-the-sdgs/
Imaz, O., & Eizagirre, A. (2020). Responsible Innovation for Sustainable Development Goals in Business: An Agenda for Cooperative Firms. *Sustainability, 12*, 6948.
Innes, J. (1990). *Knowledge and Public Policy: The Search for Meaningful Indicators* (p. 5). Transaction Publishers.
Jasanoff, S. (2016). *The Ethics of Invention: Technology and the Human Future* (p. 247). WW Norton & Company.
Jazeel, T. (2019). *Postcolonialism*. Routledge.

Jenkins, K. E., Spruit, S., Milchram, C., Höffken, J., & Taebi, B. (2020). Synthesizing Value Sensitive Design, Responsible Research and Innovation, and Energy Justice: A Conceptual Review. *Energy Research & Social Science, 69*, 101727.

Jørgensen, F. A., & Jørgensen, D. (2016). The Anthropocene as a History of Technology: Welcome to the Anthropocene: The Earth in Our Hands, Deutsches Museum, Munich. *Technology and Culture, 57*(1), 231–237.

Kaika, M. (2017). "Don't Call Me Resilient Again!" The New Urban Order as Immunology...Or What Happens When Communities Refuse to Be Vaccinated with 'Smart Cities and Indicators. *Environment and Urbanization, 20*(1), 89–102.

Kallis, G. (2018). *Degrowth*. Agenda Publishing.

Kamola, I. (2016). Situating the "Global University" in South Africa. In M. Chou, I. Kamola, & T. Pietsch (Eds.), *The Transnational Politics of Higher Education: Contesting the Global/Transforming the Local* (pp. 42–63). Routledge.

Kamola, I. (2019). *Making the World Global: U.S Universities and the Production of the Global Imaginary*. Duke University Press.

Koopman, C. (2010). Revising Foucault: The History and Critique of Modernity. *Philosophy & Social Criticism, 36*, 545–565, 557.

Kopnina, H. (2020). Education for the Future? Critical Evaluation of Education for the Sustainable Development Goals. *The Journal of Environmental Education, 51*(4), 280–291.

Kraemer-Mbula, E., Tijssen, R., Wallace, M. L., & McClean, R. (Eds.). *Transforming Research Excellence: New Ideas from the Global South*. African Minds.

Lane, R. (2019). The American Anthropocene: Economic Scarcity and Growth During the Great Acceleration. *Geoforum, 99*, 11–21, 12.

Lane, R., Warde, P., Robin, L., & Sorlin, S. (2019). *The Environment*. John Hopkins.

Langford, M., & Winkler, I. (2014). Muddying the Water? Assessing Target-Based Approaches in Development Cooperation for Water and Sanitation. *Journal of Human Development and Capabilities, 15*(2–3), 247–260.

Law, J. (2004). *After Method: Mess in Social Science Research*. Routledge.

Lea, T. (2020). *Wild Policy*. Stanford University Press.

Lehoux, P., Pacifico Silva, H., Pozelli Sabio, R., & Roncarolo, F. (2018). The Unexplored Contribution of Responsible Innovation in Health to Sustainable Development Goals. *Sustainability, 10*, 4015.

Leire, C., McCormick, K., Richter, J. L., Arnfalk, P., & Rodhe, H. (2016). Online Teaching Going Massive: Input and Outcomes. *Journal of Cleaner Production, 123*, 230–233.
Levitas, R. (2010). *The Concept of Utopia*. Peter Lang.
Liverman, D. (2018). Geographic Perspectives on Development Goals: Constructive Engagements and Critical Perspectives on the MDGs and the SDGs. *Dialogues in Human Geography, 8*(2), 168–185.
Lorde, A (1984). The Master's Tools Will Never Dismantle the Master's House. *Sister Outsider: Essays and Speeches* (pp. 110–114). Crossing Press.
Lubchenco, J. (1998). Entering the Century of the Environment: A New Social Contract for Science. *Science, 279*(5350), 491–497.
Lucas, A. (1979). *Environment and Environmental Education: Conceptual Issues and Curriculum Implications*. International Press and Publications.
Machen, R. (2018). Towards a Critical Politics of Translation: (Re)-producing Hegemonic Climate Governance. *Environment and Planning E—Nature and Space, 1*(4), 494–515.
Macrine, S. (2020). *Critical Pedagogy in Uncertain Times: Hope and Possibilities*. Palgrave Macmillan.
Marginson, S. (2010, May). The University: Punctuated by Paradox. *Academic Matters*, 14–18.
Marx, L. (2010). Technology: The Emergence of a Hazardous Concept. *Technology and Culture, 51*, 561–577.
Masco, J. (2013). *The Nuclear Borderlands: The Manhattan Project in Post-Cold War New Mexico*. Princeton University Press.
Masco, J. (2015). Terraforming Planet Earth. In E. DeLoughrey, J. Didur, & A. Carrigan (Eds.), *Global Ecologies and the Environmental Humanities: Postcolonial Approaches* (pp. 307–333). Routledge.
Massumi, B. (2018). *99 Theses on the Revaluation of Value: A Postcapitalist Manifesto*. University of Minnesota Press.
Mawdsley, E., (2021) Development Finance and the 2030 Goals. In S. Chaturvedi, H. Janus, S. Klingebiel, X. Li, A.d. Mello e Souza, E. Sidiropoulos, & D. Wehrmann (Eds.), *The Palgrave Handbook of Development Cooperation for Achieving the 2030 Agenda: Contested Collaboration*. Springer International Publishing.
Mawonde, A., & Togo, M. (2019). Implementation of SDGs at the University of South Africa. *International Journal of Sustainability in Higher Education, 20*(5), 932–950.

Mazzucato, M. (2018). *The Value of Everything: Making and Taking in the Global Economy*. Penguin.
Mazzucato, M. (2020). *Mission Economy: A Moonshot Guide to Changing Capitalism*. Allen Lane.
McCowan, T. (2019). *Higher Education for and Beyond the Sustainable Development Goals* (p. 17). Springer.
McLean, J. (2019). *Changing Digital Geographies: Technologies, Environments and People*. Springer Nature.
McNeill, J. R., & Engelke, P. (2016). *The Great Acceleration*. Harvard University Press.
Mejlgaard, N., Bouter, L. M., Gaskell, G., Kavouras, P., Allum, N., Bendtsen, A.-K., Charitidis, C. A., Claesen, N., Dierickx, K., & Domaradzka, A. (2020). Research Integrity: Nine Ways to Move from Talk to Walk. *Nature, 586*(7829), 358–360.
Melles, G., & Paixao-Barradas, S. (2019). Sustainable Design Literacy: Developing and Piloting Sulitest Design Module. In A. Chakrabarti (Ed.), *Research into Design for a Connected World* (pp. 539–549). Springer Singapore.
Merton, R. K. (1942). *The Sociology of Science: Theoretical and Empirical Investigations*. University of Chicago Press.
Miller, E. (2019). *Reimagining Livelihoods: Life Beyond Economy, Society, and Environment* (p. 88). University of Minnesota Press.
Miller, R. (2018). *Transforming the Future: Anticipation in the 21st Century*. Taylor & Francis.
Moon, C. J., Walmsley, A., & Apostolopoulos, N. (2018). Governance Implications of the UN Higher Education Sustainability Initiative. *Corporate Governance—The International Journal of Business in Society, 18*(4), 624–634.
Moore, M.-L., Riddell, D., & Vocisano, D. (2015). Scaling Out, Scaling Up, Scaling Deep: Strategies of Non-profits in Advancing Systemic Social Innovation. *Journal of Corporate Citizenship, 2015*(58), 67–84.
Morton, T. (2007). *Ecology Without Nature: Rethinking Environmental Aesthetics*. Harvard University Press.
Moseley, W. G. (2018). Geography and Engagement with UN Development Goals: Rethinking Development or Perpetuating the Status Quo? *Dialogues in Human Geography, 8*(2), 201–205.
Mula, I., Tilbury, D., Ryan, A., Mader, M., Douha, J., Mader, C., Benayas, J., Dlouhy, J., & Alba, D. (2017). Catalysing Change in Higher Education for Sustainable Development: A Review of Professional Development Initiatives for University Educators. *International Journal of Sustainability in Higher Education, 18*(5), 798–820.

Muller, F., Hoffman-Kroll, R., & Wiggering, H. (2000). Indicating Ecosystem Integrity – Theoretical Concepts and Environmental Requirements. *Ecological Modelling, 130*, 13–23.

Murphy, B. L. (2011). From Interdisciplinary to Inter-epistemological Approaches: Confronting the Challenges of Integrated Climate Change Research. *Canadian Geographer-Geographe Canadien, 55*, 490–509.

Ntuli, M. (2019, December 14). Student Activism and Its Role in Achieving the SDGs. *University World News*. Accessed Online at https://www.universityworldnews.com/post.php?story=20191214112317666

Oliveira, M. B. d. (2014). Technology and Basic Science: The Linear Model of Innovation. *Scientiae Studia, 12*, 129–146.

Orr, D. (1992). *Ecological Literacy: Education and the Transition to a Postmodern World*. State University of New York Press.

Owen, R., Macnaghten, P., & Stilgoe, J. (2012). Responsible Research and Innovation: From Science in Society to Science for Society, with Society. *Science and Public Policy, 39*, 751–760.

Pelling, M., O'brien, K., & Matyas, D. (2015). Adaptation and Transformation. *Climate Change, 133*(1), 113–127.

Pengilley, V. (2018, September 9). From Vampires to Zombies- the Monsters We Create Say a Lot About Us. *ABC Radio National*. Accessed on https://www.abc.net.au/news/2018-09-09/monsters-we-create-reflect-our-fears-and-desires/10174880

Petts, J., Owens, S., & Bulkeley, H. (2008). Crossing Boundaries: Interdisciplinarity in the Context of Urban Environments. *Geoforum, 39*, 593–601, 600.

Phoenix, C., Osborne, N. J., Redshaw, C., Moran, R., Stahl-Timmins, W., Depledge, M. H., Fleming, L. E., & Wheeler, B. W. (2013). Paradigmatic Approaches to Studying Environment and Human Health: (Forgotten) Implications for Interdisciplinary Research. *Environmental Science & Policy, 25*, 218–228.

Pietsch, T. (2016). On institutions: Why We (Still) Need Them. *Griffith Review*, Edition 51: Fixing the System, p. 117.

Plumwood, V. (1993). *Feminism and the Mastery of Nature*. Routledge.

Plumwood, V. (2013). *Environmental Culture*. Routledge.

Radcliffe, S. (2017). Decolonizing Geographical Knowledges. *Transactions of the Institute of British Geographers, 42*(3), 329–333.

Rickards, L., & Steele, W. (2019). *Towards a Sustainable Development Goals Transformation Platform*. Accessed on https://www.rmit.edu.au/research/our-research/enabling-capability-platforms/urban-futures/sdg-transformation-platform

Rickards, L., Steele, W., Kokshagina, O., & Moraes, O. (2020). *Research Impact as Ethos*. RMIT University. https://cur.org.au/project/rethinking-research-impact/

Rickards, L., & Watson, J. E. (2020). Research Is Not Immune to Climate Change. *Nature Climate Change, 10*, 180–183.

Riechmann, M., Mindt, L., & Gardiner, S. (2017). *Education for Sustainable Development Goal: Learning Objectives* (p. 10). UNESCO.

Robertson, T. (2012). Total War and the Total Environment: Fairfield Osborn, William Vogt, and the Birth of Global Ecology. *Environmental History, 17*, 336–364.

Roelvink, G. (2016). *Building Dignified Worlds: Geographies of Collective Action*. University of Minnesota Press.

Rogoff, I. (2003). From Criticism to Critique to Criticality. *Transversal Texts*. https://transversal.at/transversal/0806/rogoff1/en

Rose, D. B. (2004). *Reports from a Wild Country: Ethics for Decolonisation*. University of New South Wales Press.

Roseneil, S. (2011). Criticality, Not Paranoia: A Generative Register for Feminist Social Research. *NORA—Nordic Journal of Feminist and Gender Research, 19*, 124–131.

Ross, J., Bayne, S., & Lamb, J. (2019). Critical Approaches to Valuing Digital Education: Learning with and from the Manifesto for Teaching Online. *Digital Culture & Education, 11*(1), 22–35.

Sá, C., & Sabzalieva, E. (2018). Scientific Nationalism in a Globalizing World. In B. Cantwell, H. Coates, & R. King (Eds.), *Handbook on the Politics of Higher Education* (pp. 149–166). Elgar.

Sachs, J. D., Schmidt-Traub, G., Mazzucato, M., Messner, D., Nakicenovic, N., & Rockström, J. (2019). Six Transformations to Achieve the Sustainable Development Goals. *Nature Sustainability, 2*(9), 805–814, 805.

Salazar, J. F. (2015, July 24). Buen Vivir: South America's Rethinking of the Future We Want. *The Conversation*. Accessed on https://theconversation.com/buen-vivir-south-americas-rethinking-of-the-future-we-want-44507

Sandercock, L. (2002). Practicing Utopia: Sustaining Cities. *Disp -P The Planning Review, 38*(148), 4–9.

Santos, B. d. S. (2012). The University at a Crossroads *Human Architecture: Journal of the Sociology of Self-Knowledge, 10*(1), Article 3. Available at: http://scholarworks.umb.edu/humanarchitecture/vol10/iss1/3

Schnaiberg, A. (1980). *The Environment: From Surplus to Scarcity*. Oxford University Press.

Schneider, F., Kläy, A., Zimmermann, A. B., Buser, T., Ingalls, M., & Messerli, P. (2019). How Can Science Support the 2030 Agenda for Sustainable Development? Four Tasks to Tackle the Normative Dimension of Sustainability. *Sustainability Science, 14*, 1593–1604.

Schön, D. A. (1983). *The Reflective Practitioner: How Professionals Think in Action*. Basic Books.

Schulze-Cleven, T., & Olson, J. R. (2017). Worlds of Higher Education Transformed: Toward Varieties of Academic Capitalism. *Higher Education, 73*, 813–831.

SDSN Australia/Pacific. (2017). *Getting Started with the SDGs in Universities: A Guide for Universities, Higher Education Institutions, and the Academic Sector*. Australia, New Zealand and Pacific Edition. Sustainable Development Solutions Network—Australia/Pacific.

Sears, P. B. (1964). Ecology—A Subversive Subject. *Bioscience, 14*, 11–13.

Sedgwick, E. K. (2003). Paranoid Reading and Reparative Reading, or, You're So Paranoid, You Probably Think This Essay Is About You. In *Touching Feeling: Affect, Pedagogy, Performativity*. Duke University Press.

Shear, B. W. (2019). Toward an Ontological Politics of Collaborative Entanglement: *Teaching and Learning as Methods Assemblage*. *Collaborative Anthropologies, 12*(1), 50–75.

Shore, C., & Wright, S. (2015). Governing by Numbers: Audit Culture, Rankings and the New World Order. *Social Anthropology, 23*(1), 22–28.

Siegel, K. M., & Lima, M. G. B. (2020). When International Sustainability Frameworks Encounter Domestic Politics: The Sustainable Development Goals and Agri-Food Governance in South America. *World Development, 135*, 105053.

Siragusa, A., Vizcaino, P., Proietti, P., & Lavalle, C. (2020). *European Handbook for SDG Voluntary Local Reviews* (EUR 30067 EN). Publications Office of the European Union.

Slocum, S., Diitrov, D., & Webb, K. (2019). The Impact of Neoliberalism on Higher Education Tourism Programs: Meeting the 2030 Sustainable Development Goals with the Next Generation. *Tourism Management Perspectives, 30*, 33–42.

Sonetti, G., Brown, M., & Naboni, E. (2018). About the Triggering of UN Sustainable Development Goals and Regenerative Sustainability in Higher Education. *Sustainability, 11*(254), 1–17.

Spicer, A. (2018). *Business Bullshit* (p. 107). Routledge.

Steele, W. (2020). *Planning Wild Cities: Human-Nature Relationships in the Urban Age*. Routledge.
Steffen, W., Broadgate, W., Deutsch, L., Gaffney, O., & Ludwig, C. (2015). The Trajectory of the Anthropocene: The Great Acceleration. *The Anthropocene Review, 2*, 81–98.
Steffen, W., Leinfelder, R., Zalasiewicz, J., Waters, C. N., Williams, M., Summerhayes, C., Barnosky, A. D., Cearreta, A., Crutzen, P., Edgeworth, M., Ellis, E. C., Fairchild, I. J., Galuszka, A., Grinevald, J., Haywood, A., Ivar do Sul, J., Jeandel, C., McNeill, J. R., Odada, E., Oreskes, N., Revkin, A., Richter, D. d., Syvitski, J., Vidas, D., Wagreich, M., Wing, S. L., Wolfe, A. P., & Schellnhuber, H. J. (2016). Stratigraphic and Earth System Approaches to Defining the Anthropocene. *Earth's Future, 4*, 324–345.
Steffen, W., Sanderson, A., Tyson, P. D., Jager, J., Matson, P. A., Moore, B., Oldfield, F., et al. (2004). *Global Change and the Earth System: A Planet Under Pressure*. Springer.
Stengers, I. (2018). *Another Science Is Possible: A Manifesto for Slow Science* (p. 154). Polity.
Stephens, J. C., & Graham, A. C. (2010). Toward an Empirical Research Agenda for Sustainability in Higher Education: Exploring the Transition Management Framework. *Journal of Cleaner Production, 18*(7), 615.
Sterling, S. (2009). Sustainable Education. In D. Gray, L. Colucci-Gray, & E. Camino (Eds.), *Science, Society and Sustainability: Education and Empowerment for an Uncertain World* (pp. 105–118). Routledge.
Stirling, A. (2003). Risk, Uncertainty and Precaution: Some Instrumental Implications from the Social Sciences. In *Negotiating Environmental Change* (pp. 33–76). Edward Elgar.
Stirling, A. (2009). Participation, Precaution and Reflexive Governance for Sustainable Development. In W. N. Adger & A. Jordan (Eds.), *Governing Sustainability* (pp. 193–225). Cambridge University Press.
Straková, Z., & Cimermanová, I. (2018). Critical Thinking Development—A Necessary Step in Higher Education Transformation Towards Sustainability. *Sustainability, 10*(10), 3366.
Sultana, F. (2018). An (Other) Geographical Critique of Development and SDGs. *Dialogues in Human Geography, 8*(2), 186–190.
Szerszynski, B. (2007). The Post-ecologist Condition: Irony as Symptom and Cure. *Environmental Politics, 16*, 337–355, 351.
Taylor, C. (2019). *Posthumanism and Higher Education: Reimagining Pedagogy, Practice and Research*. Palgrave.

Thaman, K. H. (2003). Decolonizing Pacific Studies: Indigenous Perspectives, Knowledge, and Wisdom in Higher Education. Special Issue, *The Contemporary Pacific, 15*(1), 1–17.
Todd, Z. C. (2015). Indigenizing the Anthropocene. In H. Davis & E. Turpin (Eds.), *Art in the Anthropocene: Encounters Among Aesthetics, Politics, Environment and Epistemology* (pp. 241–254). Open Humanities Press.
Uluru Statement of the Heart. (2017). Accessed Online at https://fromtheheart.com.au/uluru-statement/the-statement/
Unger, R. M. (2005). *What Should the Left Propose?* Verso.
United Nations. (2020). *Sustainable Development Goals Report 2020*. United Nations.
United Nations Development Group. (2016). *Guidelines to Support Country Reporting on the Sustainable Development Goals*. Accessed Online at https://unsdg.un.org/sites/default/files/Guidelines-to-Support-Country-Reporting-on-SDGs-1.pdf
van Loon, J. (2019). *The Thinking Woman*. New South Publishing.
van Norren, D. E. (2020). The Sustainable Development Goals Viewed Through Gross National Happiness, Ubuntu, and Buen Vivir. *International Environmental Agreements: Politics, Law and Economics, 20*(3), 431–458.
Vandemoortele, J. (2018). From Simple-Minded MDGs to Muddle-Headed SDGs. *Development Studies Research, 5*, 83–89.
Waagner, H. (2011). *Meaning in Action: Interpretation and Dialogue in Policy Analysis*. M.E Sharpe Inc..
Walder, A. (2014). The Concept of Pedagogical Innovation in Higher Education. *Education Journal, 3*(3), 195–202.
Walker, J. (2007). *Economy of Nature: A Genealogy of the Concepts "Growth" and "Equilibrium" as Artefacts of Metaphorical Exchange Between the Natural and the Social Sciences*. University of Technology, Sydney.
Walker, J. (2020). *More Heat than Life: The Tangled Roots of Ecology, Energy, and Economics*. Springer Books.
Warwick, P., et al. (2019). The Pursuit of Compassionate Hope: Repurposing the University Through the Sustainable Development Goals Agenda. In *Higher Education and Hope* (pp. 113–134, 114). Springer.
Webb, D. (2013). Pedagogies of Hope. *Studies in Philosophy and Education, 32*(4), 397–414.
Weber, S. M., & Tascón, M. A. (2020). Pachamama—La Universidad del 'Buen Vivir': A First Nations Sustainability University in Latin America. In *Universities as Living Labs for Sustainable Development*. Springer.

Weller, S., & O'Neill, P. (2014). An Argument with Neoliberalism: Australia's Place in a Global Imaginary. *Dialogues in Human Geography, 4*(2), 105–130.

Whyte, K. (2016). Is It Colonial déjà vu? Indigenous Peoples and Climate Injustice. In Whyte, K. P. (2018). Indigenous Science (Fiction) for the Anthropocene: Ancestral Dystopias and Fantasies of Climate Change Crises. *Environment and Planning E: Nature and Space, 1*, 224–242.

Williams, S., & Doyon, A. (2019). Justice in Energy Transitions. *Environmental Innovation and Societal Transitions, 31*, 144–153.

Winner, L. (1978). *Autonomous Technology: Technics-Out-of-Control as a Theme in Political Thought* (p. 228). MIT Press.

Wittrock, C., Forsberg, E.-M., Pols, A., Macnaghten, P., & Ludwig, D. (2021). *Implementing Responsible Research and Innovation: Organisational and National Conditions*. Springer Nature.

Zamora-Polo, F., & Sanchez-Martin, J. (2019). Teaching for a Better World. Sustainability and Sustainable Development Goals in the Construction of a Change-Maker University. *Sustainability, 11*(15), 4224.

Zamora-Polo, F., Sanchez-Martin, J., Corrales-Serrano, M., & Espejo-Antunez, L. (2019). What Do University Students Know About Sustainable Development Goals? A Realistic Approach to the Reception of this UN Program Amongst the Youth Population. *Sustainability, 11*(13), 3533.

Zelizer, B. (2002). Finding Aids to the Past: Bearing Personal Witness to Traumatic Public Events. *Media, Culture & Society, 24*, 697–714.

Zembylas, M. (2015). 'Pedagogy of Discomfort' and Its Ethical Implications: The Tensions of Ethical Violence in Social Justice Education. *Ethics and Education, 10*(2), 163–174.

Index[1]

A
Academics/academy, v, 7, 9–13, 18, 22, 25, 28, 73–75, 79, 83, 85, 91, 93, 98, 108, 109, 114, 125, 126, 128, 129, 145, 147, 148, 153, 155, 157–160, 162–164, 171, 179, 180, 188, 190, 194, 206, 215, 228–230, 232, 236–239, 250, 251, 256, 259
Action learning, 188, 197
Activism, 6, 15, 206, 246n59
American Academy of Arts and Sciences, 81
Amin, Ash, 85, 88–90
Anthropo-Capitalocene, 251, 252, 261
Anthropocene, v, 3, 4, 10, 25, 30, 35–59, 90, 108–111, 113, 115, 120–122, 131, 132, 136, 151, 156, 161, 163, 164, 247, 249, 251, 253, 258
Anti-apartheid, 6
Arendt, Hannah, 87, 157
Attenborough, David, 1
Authentic innovation, 134–136

B
Barnett, Ronald, 71, 78, 79, 83, 84
Barry, John, 247, 251
Bayne, Sian, 193, 196
Ben-David, Joseph, 74
Biodiversity loss, v, 1, 17, 41, 234
Blades, Gen, 187
Bloch, Ernst, 253
Bohland, Jim, 76, 77
Braidotti, Rosi, 4, 195

[1] Note: Page numbers followed by 'n' refer to notes.

Brown, Eleanor, 95
Brundtland report, 56, 172
Buen vivir, 1, 25, 55, 56, 95, 138, 139, 190

C
Cameron, Jenny, 255
Campaign for a Global Curriculum of Social Solidarity Economy, 96
Capitalism, 7, 18, 39, 52–54, 74, 89, 90, 94, 99, 117, 149, 161, 163, 170, 209, 251
Capitalocene, 251
Care, 2, 10, 42, 55, 57, 69, 77, 82, 89, 90, 97, 182, 215, 217, 238, 258
Carson, Rachel, 154
The civic university, 91, 92, 94, 122, 175
Class structure, 73
Climate action, 14, 69, 235
Climate change, v, 17, 22, 25, 35, 39–41, 52, 58, 59, 69, 70, 83, 98, 116, 120, 132, 155, 156, 178, 195, 225, 227, 232, 234, 236, 237, 263
Cockpitism, 3
Colonial, 3, 10, 43–46, 48, 53, 54, 108, 115, 147, 149, 151, 235, 236, 251, 256, 257, 260
Commons, 1, 11, 16, 54, 83, 87, 91–94, 131, 160, 162, 208, 214, 223, 233, 241, 257, 258, 262
Community Economies Collective, 255
Competences, 73, 188, 189, 192, 194, 197

Connell, Raewyn, 79, 84, 85, 94, 148, 185, 228
Conscientisation, 170, 180
Consumption, 1, 2, 39, 41, 54, 58, 90, 96, 116, 126, 146, 151, 153, 156, 161, 248
Co-production, 3, 4, 14, 25, 163, 262
COVID-19, 7, 11, 18, 40, 42, 46, 52, 67, 70, 72, 77, 82, 172, 191, 226, 235, 236, 263
Crisis, v, 3, 7, 10, 11, 24, 26, 46, 52, 67, 68, 70, 72, 75–77, 82, 109, 115, 131, 157, 185, 194, 237, 239, 247, 249, 252–256, 258
Criticality, 12, 261
Critical pedagogy, 20, 28, 93, 170, 172, 175, 177, 195, 197
Critical University Studies, 74
Cultural recognition justice, 123
Curriculum mapping, 185, 215

D
Davis, Mike, 253
Davoudi, Simin, 76
De Beauvoir, Simone, 18
De Sousa Santos, Boaventura, 207, 241, 255, 256
Defuturing, 131
Democracy, 11, 16, 88, 138, 191, 228
Depoliticisation, 49
Development, v, 2, 35, 68, 108, 146, 169, 207, 247
Developmentalities, vi, 3, 7, 24, 30, 177, 235, 249, 258
Diagramming, 197
Digital divide, 191

Digital infrastructure, 191
Digital literacy, 192, 197
Disaster, 7, 111, 112, 138, 156
Dispositif, 16
Disruptive innovation, 132–134
Doughnut Economics, 259
Douglas, Mary, 49, 50
Dryzek, John, 57

E

Earth system, 39, 58, 118, 128, 151, 233
Ecocide, 3
Ecological university, 71, 78, 79, 83, 84
Economic growth, 3, 18, 25, 47, 58, 68, 70, 90, 95, 133, 137, 138, 160–162, 164, 173, 227, 251
Ecosystem, 40, 41, 77–79, 89, 90, 121, 216
Education, 3, 8, 23, 26, 28, 30, 36, 45, 70, 72, 77, 81, 82, 94–96, 108, 114, 118, 121, 131, 138, 169–179, 182, 183, 185–196, 205, 212, 222, 225, 226, 236, 249, 251, 254, 264
Education for Sustainability (EFS), 172, 175
Education for sustainable development (ESD), 28, 170, 176, 178, 184, 195, 196, 218
Energy, 13, 36, 39, 41, 154, 191, 216, 240, 248
Enlightenment, 72, 195
Environmental degradation, 39
Environmental education (EE), 172
Environmental sustainability, 8, 15, 25, 36, 70, 195, 263

Erdem, Esra, 92
Esteva, Gustavo, 54
Ethical innovation, 21, 23–27, 29, 107–139, 153, 179–182, 193, 210, 212, 249, 250, 254
Extinction, 41

F

Fear, 7, 15, 119, 149, 152, 153, 182
Fien, John, 180
1st Generation impact, 230
Food security, 48
Foucault, Michel, 16, 162
Fourth Industrial Revolution (4IR), 117, 118
Free university model, 94, 190
Freire, Paulo, 28, 93, 170, 180, 197, 253, 254
Fry, Tony, 131
Futures Literacy Laboratory (FLL), 186

G

Gandy, Matthew, 191
Genocide, 3
Gibson-Graham, J.K, 18, 34n49, 131, 255
Global, vi, 2, 6, 7, 9, 11–13, 15–17, 20, 26, 34n49, 36, 40–43, 46, 47, 51–54, 56–59, 60n11, 61n12, 67, 69, 70, 76, 82, 84, 85, 97–100, 107, 113, 125, 133, 148, 151, 152, 155, 156, 161, 170, 182, 184, 190–192, 194, 208, 221, 223, 228, 234, 236, 249, 254–256, 263, 264
Goddard, John, 91, 92, 122

Good enough universities, 86–88
The Good University, 27, 29, 83–86, 89, 94, 148, 208, 221, 228, 253
Governance by indicators, 217
Governmentalities, 75
Graduate outcomes, 184, 188–191
Grand challenges, 83, 89
Great Acceleration, 109, 111, 112, 131, 149–156, 164, 251
Greenhouse gas emissions, 40, 60n11, 69, 132, 146, 194, 222, 263

H

Haraway, Donna, 17, 119, 138
Healy, Stephen, 255
Hegarty, Katherine, 180
Higher education, v, 2, 3, 6–8, 11, 15, 24–30, 39, 43, 67–70, 74, 75, 77–81, 83, 84, 86, 91–95, 97, 99, 146, 162, 163, 169–172, 175, 177–180, 182, 183, 185, 186, 190, 191, 193, 195, 196, 198, 199n19, 206–208, 218, 219, 227–241, 248–251, 254, 255, 257, 261–265
Hillier, Jean, 197
Holdsworth, Sarah, 180
Human, 1, 4, 7, 9, 10, 12, 41, 43–45, 47, 48, 51, 55, 56, 59, 73, 87, 89, 100, 108, 110, 111, 113, 116, 117, 120, 131, 133, 138, 147, 151, 155–160, 162, 170, 175, 180, 181, 187, 195–197, 208, 209, 211, 227, 228, 241, 248, 255, 262
Humanities, 1, 69, 70, 86, 87, 147, 150, 151, 155, 160, 169, 170, 224, 241
Humanities, arts and social sciences (HASS), 129, 160
The Humboldtian Model, 72

I

Ideology, 7, 25, 54, 56, 73, 85, 95, 236, 255
Impact, v, 1, 3–5, 11, 14, 17, 20, 23, 26, 27, 29, 40–42, 59, 74, 77, 79, 82, 91, 92, 100, 107–111, 113, 114, 117, 120, 122, 126, 132, 137, 139, 146, 152, 154, 156, 157, 161, 172, 183, 184, 194, 206, 207, 213, 216, 225, 228–232, 237–241, 249–251, 255, 257, 259, 263, 264
Indicator framework, 214, 215, 218–221, 225, 235
Indigenous sovereignty, 138, 177, 235–236, 239
Industrialisation, 44, 153, 164
Innovation culture, 5, 18–23, 25, 110, 198, 210–212, 249
Institutional commitment, 5, 18, 20, 22–24, 211, 212
International development, 20, 30, 39, 42, 46–54, 56, 57, 126, 153, 216, 217, 247
International Education and Social and Solidarity Economy Network (REESS), 96
International students, 68, 81, 82

Index   289

J

Jasanoff, Sheila, 111–113, 119, 162
Jazeel, Tariq, 123
Jenkins, Kirsten, 122

K

Kaika, M., 8, 32n20, 220, 243n33
Kamola, Isaac, 6, 67, 97, 98, 239, 256
Knowledge systems, 160
Koponen, Juhani, 43, 44, 47

L

Latour, Bruno, 130
Lawrence, Jennifer, 76
Lea, Tess, 147
Learning and teaching (L&T), 6, 28, 29, 68, 86, 91, 94, 164, 169–198, 215, 218, 230, 238
Leverage points framework, 133
Levitas, Ruth, 208, 209
Library, 93, 237–239
A Life on our Planet, 1
Limits to Growth, 58, 156, 161
Liverman, Diana, 13, 217
Local knowledge, 48, 171
Lorde, Audre, 2, 264

M

Machen, Ruth, 183, 184
Mapping, 26, 97, 126, 184–186, 196, 215, 224, 225, 235
Marginson, Simon, 83, 99
Marx, Karl, 44, 130
Massive Open Online Courses (MOOC), 192, 193

Massumi, Brian, 18, 163
Matunga, Hirini, 256
Mazzucato, Mariana, 257–259
McCowan, Tristan, 17, 95
Meadows, Donella, 58, 129, 133, 138, 139
Merton, Robert, 157
Millennium Development Goals (MDGs), 36–39, 46, 53, 217, 228, 243n30, 265n7
Modernity, 13, 70, 95, 109, 171
Monsters, 6–10, 12, 32n24, 55, 149, 153
Mula, Ingrid, 183

N

Nation states, 47, 53, 180
Nature, 2, 7, 8, 28, 41, 42, 45, 49, 50, 58, 69, 70, 72–74, 83, 111, 138, 146, 162, 163, 169, 175, 187, 195, 206, 209, 215, 216, 229, 231, 233, 250
Neoliberal, 10, 15, 17, 24, 25, 50–53, 74, 76, 80, 92, 94, 95, 100, 162, 170, 171, 183, 214, 264
Newman, John Henry, 72
New normal, 67–71, 75
New social contract, 21
Non-human, 41, 95, 175, 196, 197, 228, 248

O

Online education, 192, 193
Orr, David, 169
Orr, Dominic, 180
Our Common Future, 56–58, 172

P

Paradox, v, 5, 10, 11, 13, 15, 71, 195
Paris Climate Accord, 39
Partnerships, 6, 11, 29, 74, 81, 99, 122, 214, 218, 223, 226, 229, 231, 236, 241, 264
Patel, Raj, 89, 90
Pathways, 5, 15, 22, 24, 131, 134, 135, 161, 173, 177, 184, 190, 211–213, 220, 224, 227, 228, 248, 252, 259, 260
Pedagogical innovation, 179–184
Pedagogy, 8, 14, 28, 73, 74, 85, 93, 94, 170–172, 175, 177, 180, 182–184, 186–188, 191, 195–197, 230, 236, 262
Pedagogy of the Oppressed, 170
Philanthropy, 52, 81, 149
Plumwood, Val, 90, 115, 147
Polanyi, Michael, 158, 159, 164
Policy, 30, 39, 40, 44, 49–52, 69, 74, 76, 80, 85, 109, 112, 118, 120, 147–150, 152, 158, 162, 183, 184, 217, 224, 227, 228, 231, 240, 258
Post development, 42, 53–56
Posthuman, 172, 195
Pragmatism, 252
Precarity, 11, 74, 79, 82
Precautionary principle, 118
The Principle of Hope, 253
Problem-solving, 184, 186, 188, 197, 224
Public university, v, 75, 81

R

Raworth, Kate, 31n9, 259
Reflexive modernisation, 111, 115
Reflexivity, 17, 27, 55, 110, 114, 115, 146
Regenerative innovation, 110, 136–139
Repair, 4, 57, 59, 89–90, 207
Republic of Science, 158, 159, 162
Research, v, 6–9, 14, 20, 23, 26–30, 45, 48, 68, 70, 72, 73, 75, 81, 85, 86, 90, 91, 97, 98, 108, 109, 112, 114–116, 120–123, 126, 127, 129–131, 137, 139, 142n47, 145–147, 149–153, 155–160, 163, 170, 183–185, 195, 197, 200n22, 206, 212, 215, 218, 220, 221, 224–226, 229–232, 234, 236–238, 249–251, 254, 255, 257, 263
Resilience, 4, 39, 58, 59, 62n35, 76–80, 84, 114, 138, 148, 205, 228, 234
The resilience machine, 76–80, 228
Responsible, Authentic, Disruptive, Adaptive, Regenerative (RADAR), 26, 122–123, 179, 181, 182, 250
Responsible Research and Innovation (RRI), 107, 120–123
Roelvink, Gerda, 131, 137

S

Sachs, Wolfgang, 53, 56
Schneider, Flurina, 146, 164n1
Schon, Donald, 114, 115
Science, 12, 45, 58, 72, 78, 85, 96, 112, 114, 120, 121, 129, 146, 149, 153–159, 162, 164n1, 167n39, 178, 224, 245n55, 251
Science, technology, engineering, mathematics (STEM), 160

Scientific globalism, 155
SDGs-literacy, 186, 194, 264
2nd Generation impact, 230–231
Sennett, Richard, 87, 88
Settler colonialism, 6
Shear, Boon, 171, 180
Silent Spring, 154
Sketching, 42, 197
The Slow University, 94
Social change, 11, 45, 87, 117, 184, 217, 264
Social justice, 14–16, 57, 93, 169, 173, 195, 200n26, 253, 263
Social sciences, v, 7, 15, 24, 45, 112, 132, 140n12, 150, 160, 241
Society, v, 1, 5, 7, 11–13, 16, 17, 21, 23–25, 27, 29, 30, 38–40, 45, 52, 56, 67–100, 108, 109, 111, 112, 115, 117, 118, 120, 121, 123, 130, 141n32, 146, 155, 157–159, 162, 174, 177, 186, 194, 195, 206–209, 211, 212, 216, 221, 223, 228, 232–234, 240, 241, 249, 250, 254, 262, 264, 265
Socio-technical imagination, 148
Species, 41, 147, 151
Stirling, Andy, 113, 127, 131, 140n12, 140n13
Stockholm Resilience Centre, 174
Student experience, 14, 218
Survival, 1, 44, 70
Sustainability, v, vi, 2, 5, 8, 15, 16, 25, 26, 30, 36, 42, 56–59, 69, 70, 78, 79, 94, 95, 100, 101n6, 128, 129, 134, 169–172, 174, 176, 178, 180, 183, 186, 195, 199n15, 218, 220, 223, 225, 227, 236, 237, 239, 241n2, 247, 248, 263, 265n4, 267n38
Sustainable development, vi, 2, 4, 5, 12, 23–26, 28, 30, 35–59, 89, 90, 95, 108, 110, 111, 114, 115, 117, 146, 155, 160, 164, 170, 172–178, 182, 183, 186, 192, 195, 199n13, 200n27, 207, 218, 227, 228, 237, 247–249, 251, 253–256, 260
Sustainable Development Goals (SDGs), v, vi, 2–30, 32n21, 32n22, 33n33, 35–43, 45, 46, 48–50, 52–57, 59, 69–71, 77, 83, 88, 90–92, 96, 97, 99, 100, 107, 108, 110, 113–115, 117–123, 125–127, 131–139, 145–164, 169–198, 199n16, 199n19, 200n28, 202n43, 202n47, 202n49, 205–214, 217–228, 232–241, 243n21, 243n30, 248–250, 252, 254, 256, 258, 260–265, 265n4, 266n16
Swaraj, 1
Systems thinking, 38, 133, 188, 197

T

Taylor, Carol, 195
Technological determinism, 111, 114, 118
Technology, 21, 48, 70, 107, 109–120, 122, 123, 132, 133, 154, 160, 181, 191, 193–195, 217, 224, 231
Thaman, Konai Helu, 235
3rd Generation impact, 229, 231–232

Thompson, Paul, 154
Three Horizons, 259
TIMES Higher Education Impact Index, 134
Transformation, 4, 8, 14, 18, 20, 23, 99, 118, 136, 176, 185, 193, 205, 209, 213, 226–234, 236, 240, 253, 254, 260
Trouble, 15–18, 90, 119

U

Ubuntu, 1, 95
Uluru Statement of the Heart, 2
United Nations (UN), 2, 36, 46, 47, 49, 51, 53, 177, 195, 214, 225, 240
United Nations Development Program (UNDP), 25, 35, 117
University, v, 2, 39, 67–100, 107, 145, 169, 205, 247
Unsustainability, vi, 3, 4, 10, 15, 16, 24, 30, 59, 77, 78, 247–251, 262
Utopian, 3, 71, 83–85, 208–210, 213, 253

V

Voluntary Local Review (VLR), 222–224
Voluntary National Review (VNR), 222, 225
Voluntary University Review (VUR), 214, 224–227

W

Walder, Anne, 181
Watts, Rob, 75
Welfare, 51, 52
Weller, Sally, 18
Witness statement, vi, 1, 3, 24, 30, 248, 263–265
Worldview, 25, 36, 42, 49, 50, 53, 55, 56, 95, 107, 115, 118, 119, 127, 134, 138, 139, 176, 208, 210–214, 233, 260, 261

Z

Ziai, Aram, 54, 55
Zombie, 7, 31n17, 32n24, 52, 54

GPSR Compliance
The European Union's (EU) General Product Safety Regulation (GPSR) is a set of rules that requires consumer products to be safe and our obligations to ensure this.

If you have any concerns about our products, you can contact us on

ProductSafety@springernature.com

In case Publisher is established outside the EU, the EU authorized representative is:

Springer Nature Customer Service Center GmbH
Europaplatz 3
69115 Heidelberg, Germany

www.ingramcontent.com/pod-product-compliance
Lightning Source LLC
LaVergne TN
LVHW020328260326
834688LV00037B/913